The Clean Green Home Revolution

101 Home Uses of Hydrogen Peroxide

From Toxins to Oxygen

Becky Mundt

Fifth Edition

Dedication

This book is dedicated to Mr. Walter Grotz, who took me under his wing as I wrote and refined the first edition of its publication and has taught me more about hydrogen peroxide in the ensuing years than I ever even vaguely imagined was possible. His open minded curiosity, and complete and utter dedication to the truth in spite of any 'persuasion' or even intimidation is a lesson in walking in grace. His compassionate caring and thoughtful nature has taught me far more than even his vast knowledge of hydrogen peroxide ever could. Thank you, Walter, for your willingness to talk to a complete stranger and become a lasting friend.

ISBN 978-1630221980
Fifth Edition January 2013

www.FoodGradeH2O2.com

Table of Contents

DISCLAIMER AND TERMS OF USE AGREEMENT

The author and publisher of this book and the accompanying materials have used their best efforts in preparing this book. The author and publisher make no representation or warranties with respect to the accuracy, applicability, fitness, or completeness of the contents of this book. The information contained in this book is strictly for educational purposes. Therefore, if you wish to apply ideas contained in this book, you are taking full responsibility for your actions.

The author and publisher do not warrant the performance, effectiveness or applicability of any sites listed or linked to in this book.

All links are for information purposes only and are not warranted for content, accuracy or any other implied or explicit purpose.

Foreword

November 17, 2012

It has now been six years since the first edition of this book came out in print, and it does seem that the world is beginning to get smarter about hydrogen peroxide.

What began as a personal interest in writing about a substance that has proven useful and effective while keeping our home toxin free, has developed over time and research into new discoveries and friendships.

It has blossomed into a sort of passion for learning more, not only about hydrogen peroxide but about the many ways to accomplish health, hygiene and well-being using natural substances.

Additionally, in the years since the first publication of this book, it has been gratifying to see that much of the deeper, particularly the medical, research in this area has been expanded upon and embraced by many people seeking alternative solutions to chronic health, environmental and agricultural challenges. Truly, we are only just beginning to enjoy the multitude of benefits of hydrogen peroxide in every aspect of our modern world. Particularly, I am pleased and encouraged by the work of Mr. James Roguski, who, took it upon himself to work with Walter Grotz and bring much of Walter's tirelessly collected and well documented research to a wider audience, through his website foodgrade-hydrogenperoxide.com. Mr Roguski's book "The Truth About Food Grade Hydrogen Peroxide" is a welcome and well documented account of the real medical history of hydrogen peroxide and its incredible powers of healing in the human body.

Finally, in this new edition, you'll find a new section on Poultry and Livestock, focused mainly on backyard chickens. The hot tub section has been expanded, as a result of many emails, phone calls and conversations with both hot tub owners, Mr. Grotz and others leading to new conclusions and considerations when switching a tub to hydrogen peroxide.

I am truly grateful to the many readers and website visitors who have engaged me in conversation and exploration of this fascinating topic.

Preface

Before we go diving off in to the deep end of the pool...

Just wanted to share a few thoughts with you before you find yourself sailing along in the bright new world of hydrogen peroxide.

First. Hydrogen peroxide is what it is because it is what it is. Does that sound redundant? Actually, it's not. H_2O_2, as it's chemically named, is a chemical compound. It is created in our atmosphere as a natural process, as well as in our bodies, in plants and generally all over the place.

But we humans have this capacity to take simple chemical compounds like hydrogen peroxide and make them stronger.

So all the great advice in this book IS great. Use it. Apply it. And experiment. A little.

But don't forget that you're dealing with a very powerful organic chemical compound that, in high concentrations, is an oxidizer. That means it burns. At concentrations higher than 5% it can cause permanent eye damage. So don't put it in your eye!

And respect the fact that it is strong in human created dosages. 35% hydrogen peroxide kills all microbial and bacteria life on contact in less than 20 seconds.

If you're battling the creeping crud of all mildews, that's terrific! If you're not careful how you handle it, that's not. Common sense is, unfortunately, not a salable commodity. So, just please, don't be illogical. Think science.

If you find yourself getting sloppy in your conduct, just remember: You may be playing with 8%, 10%, even 35% hydrogen peroxide. Those space shuttles are launching off the pure stuff. It IS that powerful. A little respect is not so much to ask.

1. Over 3%: Wear gloves. Don't inhale mist. Ventilate area for large applications.
2. Even at 3%: Don't inhale the mist, ventilate, let it work without working on you.
3. Any applications of eardrops, etc. never use stronger than 3% solutions.

It's like all great and good things in life. Some is wonderful, and applied correctly it gives nothing but joy. Too much makes trouble. No one thing is good all the time.

So, use hydrogen peroxide with an awareness and respect of its power, and it will serve you very very well.

Now that we've got that out of the way, let's talk about what is right, good and sensible about hydrogen peroxide. You see, that's where all the good stuff is.

Whether you are working with liquid hydrogen peroxide or sodium percarbonate dry compounds, the results and the outcomes are the same:

1. The oxidizing power will work when you apply it.
2. The same power will break down in water (any flow of water) almost instantly.
3. The end result of all actions of hydrogen peroxide will be water and oxygen, non-lethal, non-polluting, non-invasive, and not proliferating of toxic by-products in the environment.

Even if you really mess up and pour 35% hydrogen peroxide on your foot instead of into the hot tub, running water will mitigate any problem without long-term ill effects.

The inverse of the nuclear age, hydrogen peroxide, even if in concentrated form, once exposed to the natural environment will quickly, readily and reliably break down to common elements, which are non-toxic and life sustaining rather than life threatening.

Introduction

The wonders of cleaning, home and personal care with hydrogen peroxide are as surprising for their variety as for their efficacy.

In recent decades there has been an increase in awareness of the toxicity and dangers of many of the chemicals which have been introduced to our environment over the last century. This increased awareness has led to changes in industrial practices, and the identification of SuperFund sites requiring massive clean-up across the nation. Unfortunately, changes in household cleaning practices have been much less dramatic, and slower to take hold in the larger population.

New disorders, such as multiple chemical sensitivity and compromised immune response, have forced those individuals personally facing these challenging disorders to make dramatic changes to their own personal lifestyle, cleaning and personal hygiene practices.

But for those not so directly impacted, harsh cleaning agents and toxic chemicals continue to make up a large portion of the cleaning products being purchased and utilized on a daily basis.

Indeed, annual cost of household cleaning products has continued to rise. The average American family of four spends approximately $700 per year on cleaning supplies, with some variations across income levels. Generally speaking, the higher the income level, the higher the cleaning supply budget.

At the same time, research has proven there is a direct link between household indoor air quality and the types of cleaning products used in the home. The number one contributor to poor indoor air quality is the chemical composition of household cleaning products used in the home.

For those who are interested in reducing their toxic load, improving indoor air quality, reducing their polluting and harsh chemical impact on the planet and in their own immediate environment, hydrogen peroxide offers an inexpensive, convenient, and completely non-toxic alternative to literally dozens of toxic household cleaners and personal care products.

A naturally occurring compound, hydrogen peroxide is formed within the cell structures of plants and animals, in the earth's atmosphere and in the waters that cover the earth.

Formed in the upper atmosphere when water and ozone combine to produce oxygen and hydrogen peroxide, its true chemical role in the mechanism of climate and rainfall is far from being clearly understood even today.

As a commercial chemical it was first used in the restoration of famous paintings to remove sulfur buildup without damage to the paint or canvas beneath. It has been long favored in textiles pulping, fabric production and bleaching for its ability to whiten with minimal damage to fiber structures, and has been used as an environmentally sustainable replacement for chlorine bleaching in many paper and pulp operations around the world.

Hydrogen peroxide is manufactured, stored and shipped all over the world in vast quantities, yet its application as a simple and effective household chemical cleaning compound has never really come of age.

Clearly, hydrogen peroxide is a healthy, environmentally clean and sound approach to not only clean bleaching of paper pulp; cleaning of hydroponics and water garden waters, aquaculture and other industries, but is equally useful around the home.

Perhaps the first thing you'll notice once you begin using hydrogen peroxide for your household needs is a distinct lack of odor. No smells of perfumes, chlorine, cleaners, pet, or bathroom odors… Instead there will be a simple clean sweetness of additional oxygen and fresh clean air.

What else can this little wonder couple do around the home? The easiest way to understand the vast array of practical applications of hydrogen peroxide is to take a closer look at what it is, and how it interacts with the rest of the world.

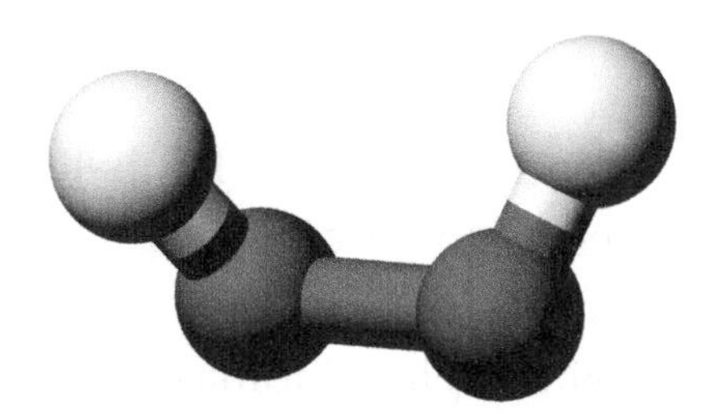

Hydrogen Peroxide: H_2O_2

Hydrogen (H) and Oxygen (O) - building blocks of the natural world.

What makes hydrogen peroxide preferable to Chlorine, Ammonia, Window Cleaner, Oven Cleaner, Pine Cleaner, Branded Disinfectants, Toilet Bowl Cleaners, and Chemical Abrasives?

Unlike chlorine, ammonia and other toxic cleaning products, hydrogen peroxide breaks down into water and oxygen. The two elements that make up hydrogen peroxide (hydrogen and oxygen) in their most common form make water (H_2O).

Environmental science recognizes classes of chemicals based on what is known as their 'environmental persistence'. Chemicals which are 'environmentally persistent' remain in the environment in their original form, or in recombinant forms created when they bond with organic matter present in the environment. These new forms are hybrids of organic compounds and the persistent chemical.

Dioxin is one such recombinant form of chlorine, and is therefore identified as an 'organochlorine'. That is to say, dioxin is formed when chlorine recombines, in certain environments, with organic matter present in that environment to create the new form, dioxin.

For a comprehensive list of the dangerous and toxic chemicals still found in household cleaning products, visit Environmental Working Group's website and review their Cleaners Database Hall of Shame:

http://foodgradeh2o2.com/ewg-hall-of-shame

You may be surprised at some of the cleaners you find on this list which you previously thought of as 'environmentally friendly', 'green' or 'non-toxic'. In many cases, cleaning ingredients are still not disclosed, even now in the 21st century! In fact, some cleaning products' warning labels will simply state such things as 'contains ingredients identified by the State of California to cause cancer' in the fine print – and no actual identification of the ingredient itself.

A Little History of Hydrogen Peroxide

Hydrogen peroxide was discovered by French Chemist Louis-Jacques Thenard in 1818. Coining the phrase "eau oxygenee" to describe its properties, Thenard believed it to be an oxygenated form of water.

Hydrogen peroxide is produced naturally within plant biomass and plays diverse and pivotal roles within the plant kingdom. It is present in trace amounts in rain, water, and snow. It is also present in higher concentrations in such natural healing springs as Lourdes, Fatima and St. Anne's.

In the late 19th and early 20th century hydrogen peroxide was the subject of many peer reviewed papers which demonstrated its efficacy against infectious disease with an extensive list of illnesses responding positively to its use. The 'peroxide of hydrogen' was made popular by Charles Marchand, whose detailed book "The Therapeutical Applications of Hydrozone and Glycozone" was published throughout the 19th century. The 18th edition was published in 1904. This book was republished by Walter Grotz in the late 1989. (In 2006, when the first edition of this book was being written, it was among the materials purchased from E.C.H.O., Mr. Grotz's nonprofit educational foundation for my library.)

Another item which was sent to me by Mr. Grotz at that time was a copy of a Lancet Journal of Medicine article from 1921 which details the first use of hydrogen peroxide in the intravenous treatment of influenza and pneumonia in 1920.

In this first use of hydrogen peroxide intravenously with influenza patients, reported in the Lancet on February 21, 1921, military doctors T.H. Oliver, B.C. Cantab, and D.V. Murphy, reported their results in treating Indian soldiers in an influenza epidemic which was claiming 80% fatality. The article, titled "Influenza - Pneumonia: The Intravenous Injection of Hydrogen

Peroxide" is still available for purchase from the Lancet to this very day.

While hydrogen peroxide has fallen out of favor in the modern allopathic and pharmacologically based medical practice, many people are beginning to become aware of and seek out hydrogen peroxide therapies, thanks to the works of people like Walter Grotz, William Campbell Douglass and Ed McCabe.

While this book does not seek to address the health care or medical uses of hydrogen peroxide, there is no question that the health benefits of hydrogen peroxide therapy are real, inexpensive and therefore extremely controversial in the age of modern big money medicine in the post John D. Rockefeller era.

More Recent History:

Interestingly, the 21st century is also full of new applications and uses for hydrogen peroxide that reach far beyond the home cleaning and health care arenas.

In January of 2007, the National Science Foundation announced a breakthrough in the remediation of disaster area polluted waters which proved able to remove deadly microbes and pathogens from Hurricane Katrina flood water samples taken in New Orleans in 2005:

"Engineers have developed a system that uses a simple water purification technique that can eliminate 100 percent of the microbes in New Orleans water samples left from Hurricane Katrina. The technique makes use of specialized resins, copper and hydrogen peroxide to purify tainted water.

The system--safer, cheaper and simpler to use than many other methods--breaks down a range of toxic chemicals. While the method cleans the water, it doesn't yet make the water drinkable. However, the method may eventually prove critical for limiting the spread of disease at disaster sites around the world."

In February of the same year, the FDA announced the approval of 35% Perox-Aid® (hydrogen peroxide) for 'control of mortality in (1) freshwater-reared finfish eggs due to saprolegniasis, (2) freshwater-reared salmonids due to bacterial gill disease, and (3) freshwater-reared coolwater finfish and channel catfish due to external columnaris disease'.

This was the first new waterborne drug approved for a disease claim for any aquatic species in more than twenty years. It is noted that the approval makes clear that ONLY 35% Perox-Aid® was approved and any other 35% hydrogen peroxide product was specifically NOT approved for this application in the aquaculture industry.

Hydrogen peroxide has continued a steady pace in the news ever since. The last few years have seen numerous news items covering the ongoing discoveries of how hydrogen peroxide works in the human body, in fish and in living tissues, the first of which appeared in Science Daily in 2009:

Science Daily reported in June 2009 that new studies showed that hydrogen peroxide plays a critical role in marshaling the immune system in defense of healthy cells.

By 2012 the news included new innovations and products designed to help the body in its natural defensive production of hydrogen peroxide (which occurs when the skin is exposed to UV light):

Excerpt: "Human skin produces hydrogen peroxide when exposed to natural light through a process called photo-oxidation. Under the supervision of Dr. Cindy Dunn, author of the clinical reference book, "Protecting Study Volunteers", UV Technologies researchers discovered the resulting hydrogen peroxide is highly effective in killing bacteria found on the surface of skin, including inside nasal passages, and ear canals. The company identified the most efficient wavelengths that trigger this response and developed UV-Aid. UV-Aid enables the user to direct this light to areas of the body that are most susceptible to disease and infection."

The other big change of the late 20th and early 21st century is the shift to hydrogen peroxide based sterilization and decontamination in hospitals, laboratories and throughout the medical establishment, as well as in remediation of toxic spills and disaster areas.

So what is it about hydrogen peroxide that makes it so effective, and what exactly does it do?

> Oxidation: An oxidizing agent is a chemical compound that readily transfers oxygen atoms. Hydrogen peroxide is a common oxidizing agent. It breaks down readily in water, becoming water and oxygen as the oxidizing agent releases its extra oxygen atom. This action of releasing the extra oxygen atom bound in the hydrogen peroxide is what is defined as oxidation.

Without water to dilute the effect, highly concentrated formulas of hydrogen peroxide are volatile, unstable, caustic and downright powerful.

It is this powerful action that makes hydrogen peroxide an exceptional rocket and space ship propellant. The simple power of oxidation when concentrated becomes capable of creating enough energy to lift spaceships off of launch pads and into outer space.

Simple yet effective, the actions of hydrogen peroxide in household concentrations (3% to 9% by volume in a distilled water solution) are also oxidative, although, by comparison, the speed with which the oxidation occurs is slowed due to the lower concentration of active ingredient. Where 35% solutions kill all bacteria and microbial life within 20 seconds of contact, a 3% solution can take up to 20 minutes to accomplish the same level of disinfection.

From Toxins to Oxygen

Replacing the Hazards in Your Home with Clean Non-Toxic Non-Polluting Hydrogen Peroxide

Toilet bowl cleaners, oven cleaners, degreasers and other chemicals used around the home can all contain extremely toxic caustic agents. Read the labels.

- Corrosives. Avoid products labeled "Danger. Corrosive." Corrosives include some of the most dangerous chemicals in the home, such as lye, hydrochloric acid, phosphoric acid, and sulfuric acid -- the active agents in many drain cleaners, oven cleaners, and toilet cleaners. These chemicals can burn the skin, cause internal burns if ingested, and explode if used incorrectly.
- Ammonia. Many home recipes and commercial products contain ammonia, but it is a strong eye and lung irritant and should particularly be avoided by anyone with asthma or other lung sensitivities.
- Bleach. For the reasons noted elsewhere, but primarily for its toxic fumes.
- Phosphates. Phosphates are naturally occurring minerals used in automatic dishwashing detergents as a water softener. When released back into the environment, phosphates can cause algae blooms in lakes and ponds that kill aquatic life. Look for phosphate-free dishwashing detergents, try a homemade recipe of half borax and half washing soda (a more alkaline form of baking soda), or skip the dishwasher and use a dishpan and regular dish soap instead.
- Petroleum products. Many surfactants (cleaning agents) are refined petroleum products that are linked with health problems and require environmentally harsh methods to extract and distill. A few specific ones to avoid: diethylene glycol, nonylphenol ethoxylate, and butyl cellosolve.

If you are in doubt about a particular product's safety, head on over to the Household Products Database from NIH and do a search for the product. You'll find their home page at: http://foodgradeh2o2.com/household-products-database-nih

Most of us consider the effects of the products we use around our home very narrowly; in terms of their effectiveness at what we want them to do, or in terms of our overall perspective on such issues as the environment, our health and the health of our family members. We have become so well entrenched in the generations-old chemical revolution which introduces new chemical products into our lives almost daily that we barely notice how many chemicals we actually use.

However, we are also now beginning, in many areas of health and environmental science, to see the evidence that proliferation of chemical toxins into our homes and environments may not always be serving us.

In the last 20 to 30 years the evidence has been mounting that a large number of the chemicals we all considered "normal household chemicals" are, in fact, highly toxic, resistant to bio-assimilation (meaning they persist in the environment after use) and become bio-toxins in our bodies and environment over time. During their proliferation beginning after World War II and right up to today, the actual environmental and health consequences of many of these chemicals have remained unknown.

Any cursory investigation of the Household Products Data Base at the National Institutes of Health cross-referenced with the Hazardous Materials Data Sheets of the Environmental Protection Agency will show that "unknown" is the most common entry in the areas of environmental and health risks of these chemicals. In fact, "insufficient research data" is the most common explanation of what the effects of these chemicals may be expected to be!

At the same time, we do know the effects of a great many chemicals, more research is being done all the time, and the results of the research that has been done are clear. Organochlorines, formaldehyde, hydrochloric acid, phosphoric acid, and sulfuric acid, phosphates, diethylene glycol, nonylphenol ethoxylate, and butyl cellosolve, and others are known toxins, which cause cancers, liver damage, reproductive disorders, immune system break down and more.

Starting with the first on the list, organochlorines make up a class of chemicals that are the result of chlorine interactions in the environment. This new class of man-made chemicals includes dioxin, and many others.

Some 177 different organochlorines have been found in fat, breast milk, blood, semen and breath in people of the U.S and Canada.

Because organochlorines persist in body fat and in the environment, they concentrate at higher levels the higher up the food chain you go.

- Some organochlorines (vinyl chloride and dichloromethane) cause mutations in genetic material, which can then give the wrong instructions to the rest of the cell for cell division, differentiation and proliferation;
- Some organochlorines (dioxin, chlorobenzenes, chlorinated pesticides, and chlorinated solvents) strengthen the ability of other chemicals to cause cancer by inducing enzymes that transform them into a more carcinogenic form;
- Organochlorines such as dioxin and PCBs interfere with the body's natural controls on cell growth and differentiation;
- Some organochlorines mimic or interfere with natural hormones like estrogen;
- Organochlorines, namely the dioxins and the solvent trichloroethan, may suppress the immune system's mechanisms for defending against tumorous cells.

Chlorine Exposure and its effects

Breathing bleach fumes, soaking in a chlorinated Jacuzzi or taking a hot shower in unfiltered chlorinated water all provide the potential for direct exposure to chlorine in a heated environment, The addition of heat to the equation causes the over exposure and added absorption of chlorine by the body.

This exposure causes two changes that affect this condition. The first effect is a CNS motor neuron-proprioception disorder whereby muscle control is disturbed, leading to differences in right and left sides or changes in antagonistic muscle function. These imbalances then translate to joints and discs, causing articular subluxations or disc shearing, with resultant nerve pressure and entrapment.

The other deleterious effect of chlorine and its chemical breakdown products is that it deranges collagen structure, changing a linear structure to a web type, circular structure, like ringworm. The structure takes on the form of microscopic scar tissue. This leads to restricted motion (myofibrositis). Eventually, the breakdown of collagen takes place in the cartilage and other connective tissues. When the support mechanism is disrupted, structural failure results, the most common of which are low back pain and sciatica.

This exposure can be prevented by eliminating household and clothes cleaning compounds containing chlorine, changing chlorinated Jacuzzi's to hydrogen peroxide sterilizing combined with ozone filtration or UV treatments and adding chlorine filtering to the household water supply, either at the point of exit (showerheads, faucet filters) or through a whole-house filtering system.

Finally, a better understanding of dioxin and other organochlorines makes it clear that while chlorine may break down into harmless salts and water in a sterile laboratory environment, in the natural environment, it does something very different than that.

The definition of dioxin from the EPA safe water drinking water hazards list includes the following:

> "What is dioxin and how is it used?"
> "Dioxin is not manufactured for any commercial uses. Rather, it is a chemical byproduct of the manufacturing of chlorine bleached paper."

The report goes on to state that "dioxin is believed to be the single most carcinogenic chemical known to science".

The U.S. Environmental Protection Agency has found dioxin to be 300,000 times more potent as a carcinogen than DDT.

In other research, scientists are beginning to assess how it is that these organochlorines make their way into human tissues.

Chlorine and its organochlorines by-products are readily absorbed by the skin. Dioxins present in bleached paper products such as coffee filters, paper towels, diapers, tissues and other products used directly on the skin may be one of the avenues by which some of the 177 known organochlorines make their way into human tissue.

It is now reported that the average American ingests a daily amount of dioxin that is already 300 to 600 times greater than the EPA's "safe" dose.

The Newest Line of Antibacterial Soaps and Why We May Want to Avoid Them

Growing evidence that a new breed of antibacterial soaps and cleaners are now having deleterious effects on North American agriculture and natural environments further suggests that we would do well to eliminate these "super soaps" from our households as well.

Why replace antibacterial soaps? Consider this excerpt from a May 2006 Los Angeles Times article:

"Tons of chemicals in antibacterial soaps used in the bathrooms and kitchens of virtually every home are being released into the environment, yet no government agency is monitoring or regulating them in water supplies or food.

About 75% of a potent bacteria-killing chemical that people flush down their drains survives treatment at sewage plants, and most of that ends up in sludge spread on farm fields, according to Johns Hopkins University research. Every year, it says, an estimated 200 tons of two compounds — triclocarban and triclosan — are applied to agricultural lands nationwide.

The findings, in a study published last week in Environmental Science & Technology, add to the growing concerns of many scientists that the Environmental Protection Agency needs to address thousands of pharmaceuticals and consumer product chemicals that wind up in the environment when they are flushed into sewers.

From dishwashing soaps to cutting boards, about 1,500 new antibacterial consumer products containing the two chemicals have been introduced into the marketplace since 2000. Some experts worry that widespread use of such products may be helping to turn some dangerous germs into "superbugs" resistant to antibiotics."

###

For many of us, the idea of thinking more globally about what we are doing locally is still a new idea. However, thinking simply of our children, and their children, it seems a natural response to want to follow a path of action that provides the greatest benefit and least potential harm to the environment we will leave to them.

Switching to simple hydrogen peroxide for household disinfecting, and general purpose cleaning is an easy, effective and environmentally sound solution.

Hydrogen Peroxide Liquid and Dry Formulations

There are two forms of hydrogen peroxide available for household use: Liquid hydrogen peroxide in various dilutions strengths, and dry hydrogen peroxide (sodium percarbonate).

Dry formulations are not active until mixed with water at 100°F or hotter. Dry formulations will break down rapidly once activated by contact with water and completely lose their oxidative powers within approximately five hours.

Liquid solutions will maintain their potency so long as they are stored properly. Both dry and liquid forms of hydrogen peroxide offer a powerful, effective, non-invasive and environmentally clean household cleaner, disinfectant, odor eliminator and all around useful chemical compound.

Liquid household and cleaning dilutions range from 3% to approximately 9%. Released oxygen molecules act on the micro-organic level to kill pathogens, molds, fungi and anaerobic life forms. In this way, hydrogen peroxide oxidation could be said to be the least toxic form of antiseptic germicidal action.

This is not to say that hydrogen peroxide is not toxic or even fatal in higher concentrations if ingested. Household strength solutions of hydrogen peroxide should be kept out of reach of children, and should be stored away from sources of heat, moisture and direct sunlight. Gloves should be worn for direct applications. Should skin become sensitive if exposed to hydrogen peroxide, the solution is always to run copious amounts of water over the affected area.

Hydrogen peroxide breaks down in UV light, when exposed to air (open container) or when mixed with water. Storage of hydrogen peroxide should be in a cool, dark location in containers which block UV light. Because hydrogen peroxide is volatile, tightly capped bottles should be loosened periodically to allow the release of built-up oxygen. Under these storage conditions, hydrogen peroxide will lose its potency only very slowly, at a rate of less than 10 percent per year.

All commercially available hydrogen peroxide is date-stamped for freshness. Over time hydrogen peroxide will lose its potency, and old containers of hydrogen peroxide should be emptied and recycled.

For many applications simple 3% hydrogen peroxide solution (same strength as found at the local drug or grocery store) will work perfectly. Particularly effective in kitchen disinfection, food grade hydrogen peroxide at

3% solution is the preferred choice for most applications where humans or food come in contact with the surfaces being cleaned.

For some cleaning jobs that require more powerful deep cleaning, bleaching and/or stain removing action, the most practical formulation is sodium percarbonate.

Sodium percarbonate, also known as sodium carbonate peroxyhydrate, or "dry hydrogen peroxide" is a mixture of soda ash, and dry stabilized hydrogen peroxide that is bound in crystalline structure to sodium carbonate. Sodium percarbonate is readily available from many environmentally friendly cleaning companies, from agricultural suppliers and directly from chemical suppliers.

Sodium Percarbonate Wikipedia definition, production and uses:

Sodium percarbonate is a chemical, an adduct of sodium carbonate and hydrogen peroxide (a perhydrate), with formula 2Na2CO3 · 3H2O2. It is a colorless, crystalline, hygroscopic and water-soluble solid.[1] It is used in some eco-friendly cleaning products and as a laboratory source of anhydrous hydrogen peroxide.

This product contains the carbonate anion, and should not be confused with sodium peroxocarbonate Na_2CO_4 or peroxodicarbonate $Na_2C_2O_6$, which contain different anions.

Structure

At room temperature, solid sodium percarbonate has the orthorhombic crystal structure, with the Cmca crystallographic space group. The structure changes to Pbca as the crystals are cooled below about -30°C.[2]

Production

Sodium percarbonate is produced industrially by reaction of sodium carbonate and hydrogen peroxide, followed by crystallization. Also, dry sodium carbonate may be treated directly with concentrated hydrogen peroxide solution. World production capacity of this compound was estimated at several hundred thousand tons for 2004.[3] It can be obtained in the laboratory by treating the two substances in aqueous solution with proper control of the pH[4] or concentrations.[2]

Uses

As an oxidizing agent, sodium percarbonate is an ingredient in a number of home and laundry cleaning products, including bleach products such as OxiClean, Tide laundry detergent,[1] and Vanish.[5] It contains no phosphorus or nitrogen. Dissolved in water, it yields a mixture of hydrogen peroxide (which eventually decomposes to water and oxygen) and sodium carbonate ("soda ash").[1]

References

[1] ^ a b c Craig W. Jones (1999). Applications of hydrogen peroxide and its derivatives. Royal Society of Chemistry. ISBN 0-85404-536-8.
[2] ^ a b R. G. Pritchard and E. Islam (2003). "Sodium percarbonate between 293 and 100 K". Acta Crystallographica Section B B59 (5): 596–605. doi:10.1107/S0108768103012291.
[3] ^ Harald Jakob, Stefan Leininger, Thomas Lehmann, Sylvia Jacobi, Sven Gutewort (2005), "Peroxo Compounds, Inorganic", Ullmann's Encyclopedia of Industrial Chemistry, Weinheim: Wiley-VCH, doi:10.1002/14356007.a19_177.pub2
[4] ^ J. M. Adams and R. G. Pritchard (1977). "The crystal structure of sodium percarbonate: an unusual layered solid". Acta Crystallographica Section B B33 (12): 3650–3653. doi:10.1107/S0567740877011790.
[5] ^ a b "Oxygen-based bleaches", The Royal Society of Chemistry, and Reckitt Benckiser (the manufacturers of Vanish)
[6] ^ McKillop, A (1995). "Sodium perborate and sodium percarbonate: Cheap, safe and versatile oxidising agents for organic synthesis". Tetrahedron 51 (22): 6145. doi:10.1016/0040-4020(95)00304-Q.

For both food grade hydrogen peroxide (as the commercial 35% solution) and quantities of sodium percarbonate larger than two pounds, federal regulations require HazMat packaging standards for shipping. For this reason it is easier and less expensive to purchase food grade hydrogen peroxide solutions at 3%, 6%, 8%, 9% or 12% solutions when buying online or where shipping will be required, or to purchase sodium percarbonate in 2lb bucket containers.

(You may purchase as many 2lb bucket containers of sodium percarbonate as you need and have them shipped together and still not be liable for

HazMat shipping containers. Hazmat shipping containers are only required for bulk packaging of sodium percarbonate (over 2lb container size).)

If you're having trouble finding sodium percarbonate or do not want to purchase it online, most agricultural and garden supply centers will either carry it, or can easily order it for you. Agricultural grade sodium percarbonate does not differ from commercial cleaning grades and all sodium percarbonate remixes to an approximate 27% hydrogen peroxide solution when mixed as a paste with hot water in a ratio of 1/4 cup sodium percarbonate to 2 tablespoons of water.

If you're thinking these are exotic or new materials that you should know more about before using, you may be surprised to know that sodium percarbonate in dilute mix with soda ash to reduce its strength to a 78% sodium percarbonate dry solid has been used in laundry and household cleaning for many years.

In most general cleaning applications sodium percarbonate is mixed in much smaller concentration amounts with hot (100°F) water. 2 to 4 ounces of sodium percarbonate to a gallon of water gives you a workable cleaning solution in the lower, 3% to 4% concentration range. For serious bleaching and cleaning of really funky items, mix a paste of the sodium percarbonate with hot water and slather on and leave overnight.

Sodium percarbonate contains hydrogen peroxide in a stabilized form, and the water must be above 100°F to activate the peroxide. Generally speaking, the hotter the better. To promote bleaching, the stronger the concentration the better.

Perhaps you have heard of a rather popular "oxygen bleaching product" which entered the home cleaning market a few years ago: "Oxyclean". Oxyclean is nothing more than sodium percarbonate and soda ash in this 78%/22% formulation!

Hydrogen peroxide, whether in its dry sodium percarbonate form or in a liquid solution, breaks down into non-toxic, environmentally friendly components: water, oxygen, and, in the case of sodium percarbonate, soda ash.

Liquid hydrogen peroxide is available in many grades and dilutions. Household uses typically range from 3% to 12%. 3% drug store grade hydrogen peroxide may be used in general cleaning applications with the understanding that it contains unidentified stabilizing agents which may be toxic. For this reason, food grade hydrogen peroxide is the preferred peroxide grade for household cleaning use.

Using Hydrogen Peroxide

We've been taught by the chemical industry to expect instant results. The price of these instant results is often toxic chemical corrosives, acids or worse. When using sodium percarbonate or liquid hydrogen peroxide solutions, it may be necessary to allow the solutions to "work" actively for a few minutes to a few hours.

When dealing with serious staining, either of concrete, carpeting, surfaces or clothing, a pre-soak period or activation time may be required. Throughout the e-book we have provided specific instructions including necessary activation time if needed for total effectiveness.

In some cases, adding a mild surfactant (a mild hand dishwashing liquid is an excellent choice) will assist the hydrogen peroxide in removing the stain. Often this is simply to help the hydrogen peroxide adhere to the stained surface in question, and/or to create a slight viscosity or "slipperiness" to the cleaning solution which aids in application.

In general, hydrogen peroxide can be safely and effectively combined with baking soda (bicarbonate of soda) to form an excellent cleaning and deep deodorizing paste. A few drops of liquid dish soap will add fragrance if desired, and viscosity.

Hydrogen peroxide should not be mixed with toxic commercial cleaners. It is a powerful oxidizing agent on its own, and without knowledge of the potential chemical reactions of toxic cleaning agents to oxidizers, it is unwise to experiment. Because hydrogen peroxide is such an effective anti-microbial, anti-fungal and anti-bacterial agent, no other chemical agents are required to accomplish proper cleaning.

The second thing about hydrogen peroxide to understand is that it is caustic at higher concentrations. The applications therefore specify which concentration of hydrogen peroxide is best suited for each particular purpose.

The human tendency to think "stronger is better" could lead to inadvertent bleaching or lightening of fabrics or other when working with hydrogen peroxide. 35% hydrogen peroxide solution is typically not suitable for direct application in cleaning processes, as it can cause oxidizing, or burning (quite literally burning holes through fabrics, for instance). A 3% to 4% solution of liquid hydrogen peroxide is perfectly effective for many household cleaning jobs; and higher concentrations are only advised for certain, specific tasks.

Rather than assuming that a stronger concentration is needed, always repeat a single application at the same concentration in order to achieve the desired result.

Amazingly, once you learn this approach, you will find that hydrogen peroxide can do almost anything around your home when it comes to cleaning, and leaves a wonderful, fresh and truly clean home in its wake.

Finally, hydrogen peroxide is a chemical compound. It is a cleaning agent, among many other things. So wear your gloves when you are getting down and dirty and digging into those chores. Your skin will not be harmed by hydrogen peroxide, but prolonged exposure can cause itching and dry the skin. Besides, who wants their bare hands in whatever is being cleaned up? For simple counter top wipe downs gloves are not necessary, but if you're headed into the boys' bathroom with a toilet brush and scrubbing tools, by all means, don the gloves!

All hydrogen peroxide containers should be clearly marked as to their dilution ratios. Most common dilutions of 3% are completely adequate for most household cleaning jobs, and in fact, this is the standard dilution we use throughout our home. In many cases, using sodium percarbonate diluted to regular cleaning strength will work quite well for more difficult to clean areas, and a stronger concentration is recommended for removing black mold, feces, urine and other odor causing organic stains. Specific recommended dilution ratios are included in each section of the book.

You will find all of the reference information you need for using hydrogen peroxide in handy at-a-glance guides and tables in the back of this book, including:

- Appendix A - Basic Cleaning Product Replacement Guide
- Appendix B - Handy reference guide and cleaning formulations table. I recommend that you print this up and keep it where it is easily accessible until you have become familiar with using hydrogen peroxide for your household cleaning.
- Appendix C - Hydrogen Peroxide Dilution Tables. Use these tables for dilution of liquid hydrogen peroxide to lower concentrations. The tables outline exactly how much distilled (or filtered, either one will do) water to add to either 8%, 12%, or 35% solutions to make the weaker solutions by volume units. There is also a chart giving the breakdown of the dilution strength of standard sodium percarbonate for varying strengths of oxidizing power.
- Appendix D - Hydrogen Peroxide Stabilizers in the Marketplace
- Appendix E - Resources

The Color Code System in This Book

Each of the 101 household uses listed in the following pages is comprised of a title and color chart for quick and easy at-a-glance peroxide concentrations recommended. The color chart correlates to the table provided in Appendix B for easy reference.

Home Use X Cleaner Strength

Examples:

General Carpet Cleaning Regular Strength
Hair Bleaching Double Strength
Wood Refinishing Extra Strength

The formulations for these three basic cleaning strengths are:

Regular Strength	3% liquid H_2O_2 or 1 oz. (1/8 cup) Sodium Percarbonate to 1 gallon water (100 to 150 °F)
Double Strength	4.5% -6% liquid H_2O_2 or 2 oz. (1/4 cup) Sodium Percarbonate to 1 gallon water (100 to 150 °F)
Extra Strength	6 – 9% liquid H_2O_2 or 3 oz. (~1/3 cup) Sodium Percarbonate to 1 gallon of water (100 to 150 °F)

Sodium Percarbonate

Sodium Percarbonate solutions will remain active for 5 to 6 hours, after which they should be discarded. Unused material may be poured down the drain. It will actually help clean and deodorize your disposal or toilet. Sodium percarbonate solutions are most effective when mixed and used in warm to hot solutions (100 to 150 °F).

Mixing Ratios for Sodium Percarbonate:

- General Cleaning:
  Mix 1 dry ounce by weight of percarbonate in a gallon of warm or hot water. 1 ounce dry weight is ~ 1/8 cup dry measurement; 2 ounces is ~ 1/4 cup dry measure.

- Heavy Cleaning/Stain Removal:
  Start with general cleaning solution first and if more strength is needed, increase dry percarbonate to 2 ounces (or 1/4 cup dry measure).
- Soaks:
  Pre-soak with 1 ounce (1/8 cup dry measure) per gallon of warm to hot water for one hour before washing.
- Paste:
  Mix 2 dry ounces (1/4 cup dry measure) of sodium percarbonate with 2 tablespoons of hot water.

> ➡ NOTE: Always wear gloves and eye protection when working with the paste strength of sodium percarbonate as it is roughly equivalent to a 27% peroxide solution in oxidizing power!

In those cases when using an extra strength formulation with Sodium Percarbonate you will mix a paste of the powder and warm/hot water to apply directly. Be sure to wear gloves if handling the paste directly. Mix the paste in a small plastic, glass or ceramic container with an old wooden spoon, painter's stick or other non-metallic utensil.

Note that the liquid H_2O_2 solutions are not listed as proportionally doubled in strength. This is because 6% and higher solutions can and will bleach some surfaces, fabrics and other materials. Always pre-test before using liquid H_2O_2 in concentrations higher than 4.5%.

Indoor Uses

General Cleaning

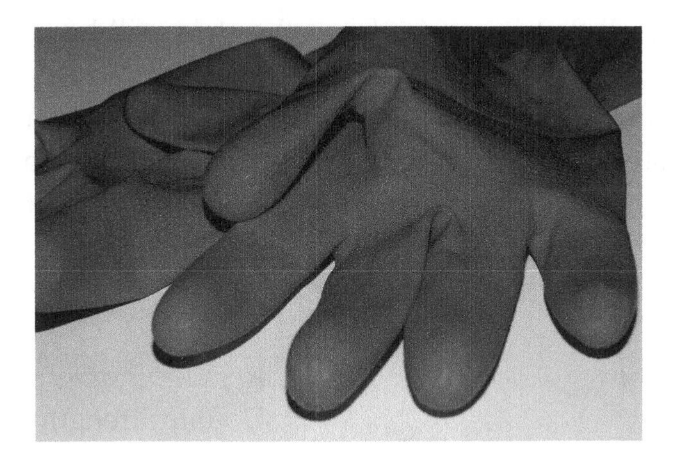

Hydrogen peroxide is an excellent all around general cleaning solution that is safe on most surfaces. A short list includes most interior surfaces and many exterior surfaces as well. From walls to windows, wood to patio concrete, laundry and carpet stains and lots in between, you'll find hydrogen peroxide is versatile and effective as a multi-purpose household cleaner. It can remove staining, mildew, mold and fungi, all without leaving harsh odors or chemical residue in its wake.

1. General Carpet and Upholstery Cleaning — Regular Strength

For general carpet cleaning to remove odors and bring a fresh scent to the carpet use 1 oz. Sodium Percarbonate or 4 to 8 oz. of 3% hydrogen peroxide added per gallon to the cleaning solution of carpet shampooers and/or steam cleaners. This solution works equally well for general upholstery cleaning.

2. Spot Carpet Stain Removal — Regular Strength or Double Strength

Remove organic stains such as grass, food, dirt or mud, tomato sauce chocolate, wine or blood. Be sure to remove all actual matter with a brush or vacuum before starting.

Spray the stained area thoroughly with 3% hydrogen peroxide and let stand a minute or two, then spray and brush or scrub with a sponge if the carpet has any nap to it to make sure the hydrogen peroxide reaches the full length of the carpet fibers. Blot with a clean cloth or towel and repeat if needed.

In many cases this will be all that is required and the stain will be gone after the first application.

For deep or seriously stubborn stains, a second application of sodium percarbonate made into a paste and left to stand on the stained area for up to 5 hours may be necessary. In this case, test for color fastness of the carpet in an inconspicuous area first.

3. Pet Odors and Stains

Regular Strength or Double Strength

Sometimes worse than the staining, is the odor pets leave behind, particularly cats. Yuck. Hydrogen peroxide is an excellent odor eliminator and works particularly well when mixed with baking soda on tough odors.

Mix a paste of 3% hydrogen peroxide and baking soda. The paste will thicken as it stands, and will re-liquefy as you stir it. Simply stir it up a bit so it's not thickened so much as to not be spreadable, and spread it over the area to be treated. Thoroughly coat the area and rub or brush the mixture into the carpet, upholstery or stained item. Let stand for five to ten minutes. In severe cases it is advisable to let stand until nearly dry, which, depending on climate conditions, can take a couple of hours.

For less severe odor problems, rinse and blot after 10 to 15 minutes with a damp cloth or sponge, continue to rinse and blot until all the baking soda residue is removed. If letting stand until dry, a vacuum cleaner may be used to remove the dry powder.

In severe pet fecal or urine problem areas we have tested pre-soaking the area in a straight 3% liquid solution, applying a baking soda top dressing which is brushed in and left overnight. Use a carpet cleaner the next day with a full 8 ounces of 3% solution added to the cleaning tank and carpet-clean as standard. This method removes stains and odor very effectively. Be sure the carpet cleaner equipment you use has excellent liquid extraction (like a good steam cleaner does) to ensure you remove as much moisture as possible when using the carpet cleaner the next day.

4. Carpet Underlayment – Especially after Water Damage

Regular Strength or Double Strength

If you have water damage that results in wet carpeting which needs to be dried out, it is a very good idea to lift up the carpet and dry it separately from the underlayment or foam padding which is typically installed under carpeting.

Whether you actually remove the entire carpet to dry out of doors, or simply lift it up where it is wet, it is a great idea to spray down the entire underlayment, or as much of it as you can get to, with 3% liquid hydrogen peroxide. This will kill the germs and bacteria that will cause the foam to become foul over time and also will kill the mold and mildew spores that will otherwise form beneath the carpet if moisture is not completely removed.

This is also worth considering for those areas where pets have stained repeatedly over time. Finding the nearest anchoring seam in the edge of the carpet and actually applying hydrogen peroxide directly to the underlayment will help prevent the return of staining moisture, bacteria and molds.

5. Black Mold Removal

Extra Strength

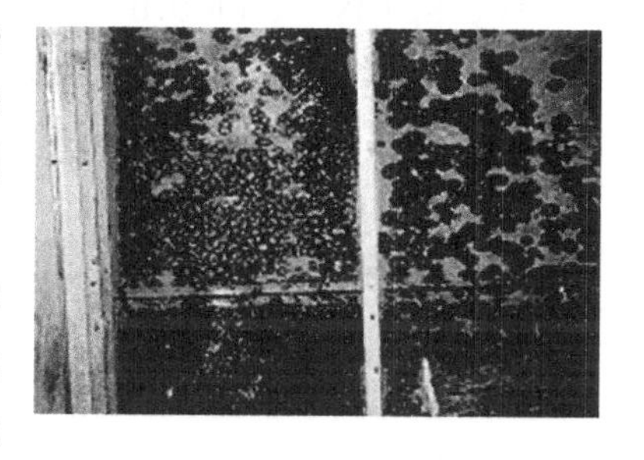

While this topic could take an entire book to cover, we'll touch on a few things that are important to know about black mold removal here.

Stachybotrys Chartarum (Toxic Black Mold) is a greenish black fungus found worldwide. It colonizes particularly well in high cellulose material, such as straw, hay, wet leaves, dry wall, carpet, wallpaper, fiberboard, ceiling tiles, and thermal insulation.

The fungus (mold), before drying, is wet and slightly slimy to the touch. There are about 15 species of Stachybotrys that can be found worldwide. This toxic mold grows in areas where the relative humidity is above 55%. Breathing in even dried mold of this kind causes many different types of respiratory problems.

Always wear a mask when removing toxic black mold. If you are unsure if the mold you are removing is toxic black mold, do some research before you begin to determine for certain what you are dealing with.

First, never ever attempt to remove black mold in a dry environment. The mold will send spores throughout the entire household if it is disturbed when it is dry to the touch. Always wet the moldy area down with liquid hydrogen peroxide spray (3 percent) or a regular strength sodium percarbonate solution before attempting to do any work at all about it.

In consultation with several mold specialists, it was recommended to use at least a 9% or 10% solution when removing toxic black mold. While 3% is fine for wetting the surface before beginning, only higher concentrations have a strong enough oxidation process to be truly effective in killing the spores completely.

Second, always close off all ventilation to the area that has mold. In basements you may need to put up plastic drop cloth sheeting to enclose the area, but do this before disturbing the mold or the mold will be spread further throughout the house rather than removed by your efforts.

In bathrooms, you can shut off all ventilation systems, close all windows and place rolled up towels at the bottom of the door and use plastic tape to seal the top of the door. Do this before beginning to remove the mold.

Thoroughly saturate the mold infested area with the 3% hydrogen peroxide spray or sodium percarbonate solution. Scrub gently with a non-metallic brush or sponge. Once the area is saturated, you can then apply the higher concentration 10% solution in spray or paste form. Continue to apply the spray or cleaning solution as you work. Make certain the area remains fully saturated as you work. A final paste of peroxide and baking soda can be applied to let sit over time and trap the mold from releasing spores as the solution dries. Wet down with fresh 3% to 10% solution when returning to clean the area after a saturation period so as to avoid ever working with a dry surface.

In severe mold cases, in stubborn corners of shower stalls, on basement walls, or in window sills with small crevices, mix a paste of straight 35% peroxide and baking soda and slather over area. The paste will adhere more easily to the area to be disinfected and hold the peroxide in place. Let stand for 24 hours and then rinse clean. Sometimes these molds have a sticking quality that will require brushing to completely remove the stain. In those cases be sure to wet down the area again with peroxide spray BEFORE scrubbing to ensure no mold spores are released into the air. ALWAYS wear gloves and eye protection when working with 35% solutions to remove stubborn molds. Molds which have been left to grow for long periods of time will usually require at least two applications to completely eliminate.

➨ NOTE: Do not be tempted to use chlorine bleach to remove mold spores from any surface. Mold spores will spread upon contact with chlorine bleach and can actually spread to new areas when chlorine bleach is applied.

6. Children's Play Areas

Regular Strength

Rather than using chemical disinfectant wipes or chlorine around children, use a spray bottle of 3 percent hydrogen peroxide. Simply spray and then wipe to clean toys, play tables, high chairs, playpens and other surfaces where children play.

You'll leave behind no toxic residues, and play areas will be bacteria, microbe, fungus and virus free.

7. Indoor Fountains

Regular Strength

Keep indoor fountains smelling sweet and fresh by adding a few ounces of 3% hydrogen peroxide periodically to the water. This will sterilize the fountain, kill algae and water born bacteria and add oxygen to the air in the room as the fountain runs.

When emptying and cleaning or setting up indoor fountains, put in one half water and one half 3% peroxide when restarting the fountain for healthy oxygenated air flow in the room.

8. Humidifiers

Regular Strength

Add 2 to 3 ounces of 3% hydrogen peroxide to humidifier reservoirs to keep mold from growing in the tank. You can add hydrogen peroxide each time you refill the tanks, or just periodically.

Between uses, or in seasons where the humidifier is not in use, keep the reservoirs dry after proper cleaning with a sodium percarbonate solution or regular 3% hydrogen peroxide rinse to keep bacteria, molds, mildew or fungi from forming in the reservoir when the humidifier is not in use.

When using larger (whole house) humidifiers with multiple tank systems or large capacity tanks, fill tanks and add ¼ to ½ cup of 3% hydrogen peroxide for every quart of water in the tank.

When caring for humidifier filters and screens, soak in a straight 3% solution when cleaning. We have successfully replaced some synthetic plastic based filters with wash cloths and other organic material based cloth which will last longer and can be washed and replaced easily. Many humidifiers come with manufactured filters or sponges which corrode, attract mold and mildew and disintegrate when used with hydrogen peroxide over time. However, our experience is that this disintegration is much slower than the time it takes for them to become unusable mold traps if no hydrogen peroxide is used.

For those lucky enough to reside in older houses which still have standing radiators which deliver hot water based heat throughout the house, it is simple to improve indoor air quality. Try the method our grandmothers used, of setting a pan on top of the radiator and filling it with water to introduce moisture into typically dry winter indoor air.

To increase the oxygen levels in the air, simply add 3% hydrogen peroxide at a rate of ¼ cup per quart of water to the pan. Do not use iron, copper or silver pans as these will act as catalysts to the peroxide and 'use it up' more quickly (as well as oxidizing your pans!). Instead, use plastic, stainless steel, glass or ceramic pans to hold the water. While aluminum is not a catalyst to the peroxide and can be used, many experts are advising that over exposure to aluminum is unwise, and so it is not a recommended material in general.

9. Walls

Regular Strength

Painted walls can be washed in a regular sodium percarbonate solution of 1 ounce sodium percarbonate to one gallon of warm to hot water, or sprayed directly with a 3 percent hydrogen peroxide solution and wiped clean.

Either way, the walls will be brighter, clean, and free of grease, dirt, mold and mildew with a simple application and wipe-clean. High traffic areas such as hall light switches will look freshly painted after the layers of dirt and fingerprints wipe-clean away.

Laminate or glossy wallpapers can also be wiped clean this way, however wallpaper should be tested in an inconspicuous place for darkening from moisture contact. Some wallpapers should not be moistened and will stain if any liquid is applied to their surface. Many of the newer wallpapers, however, can be wiped clean in this manner. Always test first, and when wiping wallpaper, spray the wiping cloth and use it to wipe the wall rather than spraying the wallpaper.

Doorjambs and molding will take on a fresh newly painted look if first vacuumed (particularly the tops of the door frames and window casings) to remove accumulated dirt and dust and then sprayed and wiped clean with a clean cloth and 3% hydrogen peroxide solution.

10. Windows

Regular Strength

Your windows will have a new bright clean shine and sparkle with no streaks when you replace your spray bottle of window cleaner with a bottle of 3% hydrogen peroxide solution. Use it just as you would any spray window cleaner: spray the window and then wipe clean with a clean cloth or paper towel. No smearing, streaking or blurring. Your glass will be clean and clear without ammonia-D, ethyl benzoic or any other harsh chemicals, which can aggravate asthma, interfere with breathing, and cause major organ damage over long term exposure.

11. Wall Degreaser before Painting

Regular Strength

Are you getting ready to paint the interior of your house? Make sure you don't have to repaint, and that the first coat goes on perfectly by wiping all the surfaces to be painted with a cloth dampened with hydrogen peroxide and let dry thoroughly, preferably overnight, before you start painting.

This will remove any grease, fingerprints or other surface matter that prevents the paint from adhering to a clean surface.

Kitchen

The Kitchen uses of hydrogen peroxide can replace a whole shelf of cleaners in your closet. From general cleaning, produce, egg, cheese and meat washing to scouring and surface cleaning, hydrogen peroxide can literally do it all.

In 1997 food scientist Susan Sumner, at the Virginia Polytechnic Institute and State University, developed a chlorine-free home regimen for disinfection that has proved to be very versatile; using simple spray bottles of hydrogen peroxide at 3% solution and white vinegar. Her studies showed that this combination not only killed all microbes associated with contaminated red meat, but also all microbes on metal, wood and plastic kitchen surfaces. The same formula kills microbes found on the foods themselves.

Spray vegetables first with a 3% hydrogen peroxide solution and then follow with a white vinegar spray. In fact, as Ms. Sumner found, the order of which solution is sprayed first does not matter.

The solutions represent an adaptation of a chlorine free disinfection scheme Ms. Sumner had been working on for red meat, and which turned out to be effective for decontaminating carcasses. In the course of her studies, Sumner found that **vegetables coming from the garden or farm not only tend to bear far more germs than red meat does, but they also hold onto germs more tenaciously.**

While most germs that show up on produce come from the soil and are benign, reports of Shigella on cantaloupe and Salmonella on raw vegetables prompted Ms. Sumner to develop **a bactericidal treatment for restaurants** and other purveyors of salads.

In her tests, she deliberately contaminated clean fruits and vegetables with **Salmonella, Shigella, or E. coli O157:H7**, all capable of inducing gut-wrenching gastroenteritis. On its own, the hydrogen peroxide was fairly effective against all three germs, she found. But the best results came from pairing the two. **"If the acetic acid (vinegar) got rid of 100 organisms, the hydrogen peroxide would get rid of 10,000, and the two together would get rid of 100,000."**

In a report by Online Science News Ms. Sumner was quoted as saying: **"What I really liked about this treatment is that every [microbe] that drips off is killed."** That means you are not just transferring disease-causing contamination from your food to the sink, drain, or cutting board. Speaking of which, she notes that the paired sprays work well in sanitizing counters and other food preparation surfaces, including wood cutting boards.

What about taste? The peroxide left no lingering flavor, and the vinegar, when applied to the skins of such vegetables as tomatoes and peppers, was undetectable; it was possible to discern the slight taste of vinegar on lettuce leaves. However, a quick fresh water rinse followed by gentle shaking or run through a salad spinner easily removes any lingering taste. For salads eventually dressed in vinaigrette, the simple rinsing is not even necessary.

12. Fruit and Vegetable Bath

Regular Strength

Pesticides, bacteria, pathogens, chemicals, dirt and general grime can all be removed safely and easily from all our fruits and vegetables with a simple veggie bath. Fill the clean kitchen sink half full with cold water. For most of us that's about 4 gallons of water. If your sink is unusually large or small, simply use a milk container and measure how many gallons equal one half of your sink in water.)

Add ½ oz. of 35% hydrogen peroxide or 5 oz. of 3% hydrogen peroxide. Place fruits and vegetables in the "bath" according to their type: lettuces and leaf veggies 15 minutes; apples, cucumbers, celeries, squashes 30 minutes. Rinse, dry, and refrigerate after the bath. If I'm in a hurry when I get home, I simply spray them all down with 3% solution followed by a spray of white vinegar, set to air dry and then refrigerate.

When returning open heads of lettuce or leafy greens to the fridge, a quick spray with 3% hydrogen peroxide keeps them fresh and green, preventing brown rot on the leaves. Be careful not to saturate the leaves, just a quick misting will do. If you accidentally over soak them to the point of dripping wet, pat off the excess with a clean cloth before refrigerating.

13. Salad Spray

Regular Strength

When preparing salads in advance of a meal, mix all ingredients except the dressing, toss and spritz gently with 3% hydrogen peroxide before covering and refrigerating. The salad will be fresh and crisp at the table and the only thing you'll have added to the mix is a little oxygen and fresh water!

➡ Note: a "gentle spritz" is a **few sprays at most**, do not saturate the salad as your aim is merely to moisten the air exposed surface of the salad with hydrogen peroxide and provide that oxygen barrier that keeps the salad fresh!

14. Fresh Sprout Growers

Regular Strength

Fresh alfalfa, mung bean, adzuki bean, broccoli, sunflower and clover sprouts are extremely nutritious and easy to grow. Delicious in sandwiches and salads, they quickly become a household favorite once you try growing them.

We also sprout mung beans and adzuki beans as a special treat for our chickens, because it is so simple, cost effective and time and energy efficient and the chickens go completely bonkers over them. It's fun to give them a little excitement now and then. And the high nutrient value of those sprouts goes right into the eggs they give us in return.

Recent concern over commercially grown sprouts and other organic greens make it an even better idea to "grow your own" when it comes to sprouts, and, frankly, nothing could be easier, especially if you have hydrogen peroxide to help you along.

Ensure your sprout growing success by starting with washing the sprouting equipment (sprout trays, plastic planters, glass jars, etc.) with regular hand dish washing liquid and hot water. Next, spray the interior surfaces with 3% food grade hydrogen peroxide solution, followed by a spray down with white vinegar. Rinse with clean water, then wipe the trays until clean and dry with paper towel or a clean cloth. This will disinfect the growing area before you begin.

Next, follow the instructions for your sprouting seeds, either by setting them in a glass jar with clean water to soak for a bit, or by moistening and placing in the planting trays. However, instead of using plain water to rinse and prepare your sprout seeds, use a hydrogen peroxide dilution of 4 ounces of 3% food grade hydrogen peroxide to one quart of water.

Each time you rinse the sprouts or water them in the growing cycle, use the premixed hydrogen peroxide solution instead of plain water.

Your sprouts will be fresh, super green, and delicious, and best of all they will not become subject to mold and mildew as can happen with home grown sprouts.

➡ Tip: Be certain to rinse and/or water growing sprouts at least once a day, preferably twice a day during the growing cycle. This prevents them from drying out, and also re-sterilizes their surfaces to prevent any wild yeasts, molds or fungi from adhering to their surfaces.

➡ Tip: remember to repeat the washing and disinfecting steps each time you start a new crop of sprouts. This way your sprouting trays cannot harbor molds or other airborne yeasts, mildews and fungi.

15. Prevent Mold on Strawberries

Regular Strength

Nothing ruins a great batch of strawberries faster than that pesky grey greenish mold that creeps up from the bottom where you don't see it and makes strawberries inedible. Yuck.

Don't let it ruin your strawberries. From now on, give those strawberries two things right away when you get them home: a quick swishing through a hydrogen peroxide bath and then a nice airy spot to dry. If you are in a rush, pat them dry with a clean cloth or paper towel. Then put them away. They'll keep better, taste sweeter and best of all, no yucky mold will grow and ruin them.

(Of course, if you leave them for weeks at a time, there is no telling what will eventually grow on them; I am only speaking here of using them within a few days to a week. This is not a long-term science experiment! Although perhaps, now that I've thought of that I'll do a test and tell you just how long I can get them to last… ☺)

16. Meat Sanitizer/Disinfectant

Regular Strength

Meat is contaminated by handling. This is why in the large slaughterhouses human hands are never supposed to come in contact with the meat. Of course, in this modern age of factory farmed meats, the meat coming from such facilities is so over-exposed to antibiotics, ammonia products (even ammonia sprays!) that we simply cannot force ourselves to buy any of it.

In small local shops the meat will be cleaner to start with, but may still require that you take precautions. Once you get the meat home, the same rules apply: handling can cause contamination. This is particularly true if you are handling different raw meats at the same time (chicken, beef, pork, etc.)

Before you begin to work with your meat or poultry, spray your work surface with hydrogen peroxide, followed by a spray of white vinegar. Then spray down the meat or poultry itself. Let stand a few minutes before patting dry with paper towel and beginning preparation for cooking.

Meats should be kept refrigerated before use, however, for best results meats should be allowed to reach room temperature before cooking. Simply follow the steps above an hour or two before you plan to cook the meats, cover lightly with clean plastic wrap or in a marinade container such as a Tupperware, and your meats will be tender, juicy and perfect without risk of contamination.

17. Marinade/Meat Tenderizer

Regular Strength

Hydrogen peroxide is an excellent meat tenderizer for poultry. For best results, spray the meats with a light misting of hydrogen peroxide early in the day and let the meat rest in the refrigerator until an hour or two before you intend to use it.

There are those who suggest submerging the poultry in a 3% solution overnight, but this is not a method I've employed yet.

Then season, prepare and grill, roast or cook as normal. You'll get rave reviews for your beautifully tender meats.

18. General Surface Cleaner

Regular Strength

This leads us right to those surfaces you're working on. Clean them up with hydrogen peroxide and white vinegar and you won't be worrying about soap residues OR bacteria and you won't have to worry about the smells or tastes of cleaners in your food.

From cutting boards to counter tops, this powerful duo will ensure you have a clean sterile surface.

Counter tops, stovetops, even the walls are safely, effectively and easily cleaned with a simple one-two spray of white vinegar and hydrogen peroxide. If you have particularly stubborn cooked-on food or dried food on stove top or counter surfaces, simply spray and let sit a bit first. The solution will quickly begin to dissolve the food residue and it will easily wipe clean after only a few minutes.

You can always tell when the hydrogen peroxide is working, because you can see it and hear it, so take a look and listen, and if it's still stuck (I doubt it), spray again and wait a moment.

19. Floor Washing

Regular Strength

Speaking of surfaces, don't stop at the counters when it comes to using hydrogen peroxide. I use it all the time to wash the kitchen floor. The great thing is that it works just like a "spray and mop" solution... just spray, and mop! It leaves your floors really clean, so you don't have that sticky soapy after-effect that the commercial spray and mop products leave behind. Plus, it disinfects as it goes, so your kitchen floor could actually be clean enough to eat off of! Just kidding, don't be putting food on your kitchen floor, really!

20. Degreaser

Regular Strength or Double Strength

Hydrogen peroxide and white vinegar makes the best degreaser I've ever used. Spray greasy surfaces directly with the vinegar, then spray the hydrogen peroxide and, depending how thick the grease is, wipe immediately or wait a few minutes to let the peroxide do its work. To clean the stove exhaust hood, place newspaper or paper towel below the hood on the stove or counters to catch the grease as it runs off the hood. Spray down thoroughly, then wipe clean. Repeat until all the grease is removed.

This even works in homes where we came in after no one had cleaned the stove hood in what appeared to be years. It just takes more newspaper and a few more applications of the process, but it leaves a squeaky clean grease free surface.

21. General Disinfectant/Sterilizer/Germicide Regular Strength

Cutting boards, knives and surfaces where you prepare raw meats, vegetables and other foods can be kept bacteria free with simplicity using your hydrogen peroxide and vinegar spray bottles.

Spray and then wipe down the surfaces after use to avoid spreading the bacteria through the kitchen. Particularly when preparing raw meats and vegetables for cooking, it is important to remove the packaging, any meat drippings and other potential bacteria sources right away.

Because vinegar and hydrogen peroxide kill all microbes and bacteria that can collect in any food area, it's a good idea to use them regularly in your entire kitchen cleaning. You don't have to worry about chemicals, toxic cleansers or soap residue and your kitchen will be fresh and clean all the time.

22. Coffee Maker Cleaning Regular Strength

Do you use a coffee maker for your daily cup of brew? It is surely a wonder of modern times to have a machine that can have that cup of coffee hot, ready and waiting for you when you awaken each morning. But cleaning those machines can be a real chore. Not cleaning the machine means the coffee starts to taste bitter or worse, and that is simply not an option. Instead of running cleaners through the coffee maker, simply fill the water chamber halfway with water, add vinegar to ¾ full and hydrogen peroxide to complete filling the container. Then run the machine as usual - sans coffee. If your machine uses paper filters, you will want to place one in the machine before starting to catch the residue that will come out of the machine as it cleans.

After the mixture has run through the machine, wipe all the interior surfaces of the machine you can get to with a cloth sprayed with 3% peroxide, to remove the built-up residues underneath the filter apparatus and in the filter compartment. Then run the machine again, this time using only water and ½ cup hydrogen peroxide to remove the vinegar residue and the last of the crud loosened by your wiping of interior surfaces.

Last, simply run a few pots full of fresh clean water through the machine, and voila, your coffee will taste fresh and wonderful once more. Between cleanings, run a damp cloth or clean sponge sprayed with peroxide under the filtering apparatus occasionally to remove the buildup of coffee residues.

➡ Note: if your machine has a chlorine filter insert or floating filter for the water chamber, remove it before cleaning the machine.

23. Drain Cleaning

Regular Strength

Not only is hydrogen peroxide and sodium percarbonate great for your pipes (plumbing pipes that is), but also it's excellent for clearing drains, and keeping the septic system healthy.

For general drain clearing use a regular cleaning solution of sodium percarbonate and very hot water and pour the mixture directly down the drain. Do not run water for at least a few hours, or if possible, overnight. For really clogged drains, make up an extra strength solution (3 to 4 ounces of sodium percarbonate per gallon) and again use very hot water. Mix the solution for a minute or so. Pour 1 to 2 cups of white vinegar down the drain, followed immediately by the hot sodium percarbonate solution.

To improve the health of septic systems, add one gallon of regular strength sodium percarbonate cleaning solution down your drainpipes every few months or as needed to maintain freshness and looseness of soil in the drainage field.

24. Dishwasher Additive (Sanitize / Germicidal)

Regular Strength

You can use sodium percarbonate directly in the dishwasher. You may add sal soda, soda ash or baking soda if desired; however it is not necessary. Add ½ oz. per dishwasher load along with 1/3 the normal amount of dishwashing soap for best results.

➡ Note: Do NOT add sodium percarbonate or hydrogen peroxide to dishwasher loads which contain silver or copper items in the load. Silver and copper are catalysts for peroxide, speeding up the oxidizing process and they will leave a white, dulling powdery effect on the silver as it is oxidized by the peroxide. So, when washing silver, skip the peroxide!

25. Refrigerator Cleaning

Regular Strength

For many years we have been told to clean our refrigerators with chlorine bleach. This is definitely not a good idea. Chlorine bleach breaks down on contact with organic matter to form "Organochlorines" which are a new class of chemicals now being recognized as extremely persistent in the environment as well as being extremely toxic. Dioxin is just one of hundreds of these new chlorine compounds; some of which are now known to be thousands of times more carcinogenic than DDT. Never clean your refrigerator with chlorine bleach.

For a fresh, clean and environmentally sound and healthy refrigerator, start by removing all the contents of the fridge. Spray the entire interior with white vinegar followed by hydrogen peroxide. Be sure to spray down the walls and ceiling of the fridge as well as the shelves, as harmful molds, mildews and bacteria can form on these surfaces as well. Wipe clean with a sponge or clean rag.

Vegetable, fruit and meat or cheese drawers often get funky odors over time. Take them out, spray them down inside and out and let them stand in the fresh air while the peroxide does its work to get them fresh again. Then wipe down and replace.

When cleaning old or long un-used refrigerators, you may wish to soak removable shelves and drawers in a hot water bath with a regular strength solution of sodium percarbonate before proceeding to the above method.

26. Sink Bleaching

Double Strength or
Extra Strength

For white porcelain sinks hydrogen peroxide is naturally brightening and whitening every time you use it in the sink. To get a super white bleaching, spray the sink with 6% or 9% solution and let stand overnight before rinsing.

Using a 3% solution regularly will brighten the sink but for actual bleaching to occur you will need a higher concentration solution.

You can also mix a paste of sodium percarbonate and water to spread on the sink as you would a scouring powder. Let stand 10 minutes to several hours, and rinse clean. Remember that sodium percarbonate in such a paste form will be an approximate 27% solution, so wear gloves if applying with anything other than a brush with a handle that prevents direct skin contact with the paste.

27. Tupperware Stain Removal/Refreshing — Regular Strength

Is your Tupperware looking less than stellar? There is nothing like a stained and stinky Tupperware to turn you off from storing anything, let alone food you plan to eat later! Whether from spaghetti sauces, heavy garlic or onion based foods or strong staining spices and seasonings, Tupperware can become unsightly and stubborn to clean over time. To get rid of those stains and the smells they carry, soak your Tupperware in a hot sink of soapy water mixed with white vinegar and hydrogen peroxide. For instant cleaning, spraying hydrogen peroxide directly on the Tupperware also works very well. The Tupperware will look like new and feel and smell clean and ready to use.

The first time I tried this I'd been trying to clean a large Tupperware stained with spaghetti sauce using regular dishwashing liquid. It simply wasn't working. I grabbed my trusty spray bottle of hydrogen peroxide and sprayed down the Tupperware inside and out… and as I continued washing, it instantly came completely clean. I was hooked.

Your Tupperware containers will last longer and never cause your food to taste "off" again.

Scouring powder for tough grime or cooked-on foods – mix one part baking soda with one part 3% hydrogen peroxide. Mix together to form a paste and cover the area to be cleaned with the paste. Let stand a few minutes, then wipe clean. For very stubborn stains you may have to repeat the application.

Glass and chrome appliances all clean up beautifully with straight hydrogen peroxide 3% solution without leaving streaks, grease or residue behind.

28. Plastic Utensils and Picnic Dishes

Regular Strength

Sterilize and clean outdoor plastic ware with a regular solution of sodium percarbonate. Simply wash the items in the sodium percarbonate cleaning solution, rinse, and dry. For stuck on or stubborn stains, let the items soak up to one hour before washing and rinsing. Clean, sterile and ready for the next use.

29. Kitchen Sponge Life Extension

Regular Strength

Keep sponges and dishcloths fresh and clean by spritzing them with the same one-two sprays of vinegar and peroxide between uses. Start with the vinegar first if you do not like the odor of vinegar, and spray the peroxide last.

Sponges can be soaked in a cold water hydrogen peroxide and salt solution. Let soak overnight. In a bowl combine ½ cup hydrogen peroxide and ¼ cup salt. Swish to dissolve the salt, and place sponge directly in the mixture. Soak overnight and simply rinse and squeeze the sponge dry in the morning.

To prolong sponge life between soakings, a good spray of 3% hydrogen peroxide solution allowed to sit in the sponge a few minutes before rinsing out works wonders.

30. Oven Cleaners

Double Strength or Extra Strength

Oven cleaners are among the most toxic of all household cleaners. They contain volatile acids and toxic chemicals, which can damage the liver, kidneys and respiratory system. Asthma sufferers should never be exposed to the fumes from such toxic cleaners, but then, neither should anyone else.

To avoid the whole problem of toxic oven cleaners, the first step is to clean the oven regularly. This will create much less work than allowing spilled food and liquids to build up and carbonize over time. If your oven has a heat cycle cleaning unit, you may only wish to clean using that system, but even after a heat cleaning, wiping the oven out with a clean cloth sprayed in 3% hydrogen peroxide solution will pick up all the remaining soot, dirt and ash from the heat cleaning cycle.

To clean the oven without a heat cleaning cycle, sprinkle baking soda (approximately ¼ to ½ inch deep) over the major spills encrusted on the oven floor and saturate with 3% hydrogen peroxide. Finally, spray the area with vinegar. Let stand approximately 30 minutes, or up to 1 or 2 hours for more stubborn thicker masses.

If the solution has become dry while standing, see if it will simply brush clean with a cloth, often it will. If not, re-wet the area with hydrogen peroxide and then use a sturdy cleaning pad to remove the stain.

For extremely stubborn built up food/carbonized spills which have set a long time in the oven floor, you may need to start with a mixture of vinegar and baking soda in a paste and then spray on the hydrogen peroxide to add the oxidizing action to the foaming acid/base reaction of the vinegar and baking soda. This can take time to do its work and should be left to stand until all fizzing and bubbling has ceased. Then use a plastic spatula to scrape the loosened crud and remove it and follow the steps above for regular cleaning.

Occasionally, someone will complain to me that this is not as easy or as fast as using a commercial oven cleaner. I try to remind them that toxic chemicals may work more quickly and without any effort, but they are toxic. I remember the veterinarian who was explaining that natural safe remedies for fleas would not kill all fleas instantly on contact. That is because these products are not toxic and deadly to either the fleas or the pets, and require more patience and time to work effectively. The same is true with oven cleaning.

31. Toaster Oven Cleaning

Regular Strength

Toaster oven cleaning is very much the same as regular oven cleaning. Unplug the appliance before starting clean up! Remove all the loose crumbs and debris from the bottom tray of the toaster oven. Apply baking soda and hydrogen peroxide to the stubborn stains, spray with white vinegar and let sit. Then wipe away for a clean finish.

To remove grease or sticky messes from the exterior of the toaster oven, simply spray with hydrogen peroxide and wipe clean. Works great on those little glass windows in the toaster oven door too!

32. Microwave Cleaning

Regular Strength

Microwave ovens come sparkling clean with hydrogen peroxide. It's a snap to do as well. First, remove the glass plate and its stand from the bottom of the oven and wash them separately in the sink. You can wash them with dish soap or use the peroxide for this job as well, and if you do wash them with dish soap rinse and then spray with the peroxide to remove all odor causing bacteria and to disinfect them.

Then, simply spray the entire interior surface of the microwave. Do not spray directly onto hot surfaces! If the microwave has just been used, you must wait until the interior is cool to clean it. Let the peroxide sit for a moment and then wipe clean with a clean sponge or cloth. Here too, the front glass and interior panels will come easily and quickly clean, free of grease, food particles and other debris.

Another great benefit of this cleaning method is that it eliminates all odors that accumulate in microwave ovens. There is something awful about heating a cup of tea or coffee in a microwave which smells strongly of exotic or garlic-heavy food. A quick spray-down and wipe of the interior of the microwave even when it doesn't need a full on cleaning will immediately eliminate all those odors.

> ➡ Tip: Always use a soft cloth or non-abrasive sponge when cleaning the interior of the microwave to avoid scratching the interior surfaces.

33. Ant Control

Regular Strength

Every year in late summer we have a fresh invasion of ants. Seems that when the weather has been dry long enough, they come in looking for water. The kitchen can be ant free one day and overrun the next.

The first thing to do is figure out where the ants are coming in to the house. In our case it always seems to be either the kitchen window over the sink or the back kitchen doorsill. Wherever it is, find the trail where it enters the house.

In the first step of this process, we put out a small dish (a tiny ramekin, or even an old tuna can works just fine). Place either a small amount of diet soda containing Equal/Aspartame or a small amount of water with a packet of Equal diluted in the water. Set the dish along the ant trail close to where the ants are congregating. Usually this will be near the sink, or on a counter that leads to the sink.

Next, spray down the remaining ant trail indoors, as well as all surfaces where the ants have been, with 3% hydrogen peroxide and wipe clean. Use a sponge or rag, rinse and wipe until you have killed all the ants and the path of the ant trail is all wiped. The hydrogen peroxide will destroy the ants' scent trail and they will no longer continue to enter the house following the trail.

Within about 24 hours you will notice that any ants still remaining in the house will be in and around the dish of diet soda or sweetener and water, and most, if not all of them, will be dead. Remove the dish, and wipe the counters and surfaces clean once more with hydrogen peroxide.

The ants will not return. In our kitchen we keep a ceramic honey jar on the counter, in past years with bad ant infestation we had to keep the honey in the fridge to avoid re-infestation. Now that we've switched to this method, we never have to put the honey in the fridge (where it gets too stiff and hard in the cold for easy use) and the ants never return after the first application of this method.

To continue to keep your household ant free, routinely spray and wipe down all food surfaces with hydrogen peroxide after use. In our house, we simply wipe down the counter where the honey jar is stored each night. No more ants and no more cold stiff honey either!

34. Fruit Fly Control

Regular Strength

Another common kitchen pest in summer is the fruit fly. These tiny flying insects are capable of breeding at super-fast rates and attacking any open fruit or vegetable in the kitchen. Suddenly your fresh fruit basket is lost in a swarm of these little pests who are stinging the fruit to lay their eggs. Yuck.

The first line of defense is to wash or bathe the fresh fruit and veggies when you bring them home, especially those that you keep at room temperature in the kitchen. This will protect the outer layer of the fruit and reduce the incidence of fruit flies generally.

Next, if you do discover fruit flies in your kitchen, spray them with hydrogen peroxide wherever they are. It will knock them out of the air for easy wiping up without putting any harmful pesticides on or near your food. Keep compost containers and other food sources tightly sealed and wipe down their outer surfaces with hydrogen peroxide solution to remove any odor and eliminate the bacteria that attract the fruit flies.

This method has worked very well for us to eliminate fruit flies when they do appear and to keep the incidence of their appearances much lower.

Bath

35. Toothbrushes

Regular Strength

We use our toothbrushes usually several times a day. We rinse them out with water, and place them back in their rack, or in a cup on the sink side. However we do it, most of us don't think much about it until our toothbrushes start to get old, lose bristles or simply get that "smashed" look to them... and then we replace them.

But how can you really keep your toothbrush clean? A great way is to put about an inch of 3% hydrogen peroxide solution in a cup and drop your toothbrush in on its head. Let it sit there until the next time you use it. Then dump the solution, rinse out the cup and the brush in clean water and you have a completely clean disinfected and bacteria free toothbrush. It won't make the brush last longer perhaps, but it will make you feel like using it more often!!

> ➡ Note: 3% hydrogen peroxide will destroy all microbes, bacteria, fungi and pathogens in approximately 20 minutes. Any time you want certain disinfection, simply let the toothbrush soak for no less than 20 minutes in the 3% solution.

36. Hairbrushes

Regular Strength

For most brushes, a good cleaning is begun by removing all the hair in the brush and then running the brush under hot water or swishing in a sink of dish detergent or shampoo and then rinsing and allowing to dry. Next time you wash your hairbrush, try adding a few ounces of hydrogen peroxide to the sink along with whatever mild soap you usually use. Your brush will be cleaner, and any mites or other tiny organisms which live on our skin and hair will be eliminated.

37. Bathroom Cleaning General

Regular Strength

A spray bottle of hydrogen peroxide in the bathroom is perfect for general cleaning. From walls to windows and windowsills, toilet tops to mirrored cabinets, hydrogen peroxide leaves a residue free clean surface behind.

Remember to allow the peroxide to work when cleaning bathrooms that really need it – don't wipe immediately after spraying down grimy surfaces. Instead, spray and let stand for up to 20 minutes and then return, spray again and wipe down.

38. Replacing the Bathroom Toxins: Acids, Chlorine, and Ammonia

Regular Strength

Baking soda and hydrogen peroxide will do a very good job of removing hard water stains, scale and scum, and leaves white surfaces super white and bright without the risk of bleaching on colored surfaces. Use the baking soda and 3% liquid solution in place of sodium percarbonate paste when bleaching is not desired but brightening is. Remember that sodium percarbonate in a paste is approximately a 27% cleaning strength solution of peroxide so it CAN and likely will cause bleaching. This is why the baking soda and 3% solution is the alternative when bleaching is not desired.

39. Soap Scum Buildup Prevention

Regular Strength

One of the best and easiest ways to reduce soap scum build up in the shower and tub is to spray the interior surfaces of the shower cabinet and/or bath with hydrogen peroxide after each use. Makes that routine bathroom cleaning a LOT easier!

40. Grout Stain Removal

Double Strength or Extra Strength

Cleaning tile grout of black dirt and mildew or discoloration is easiest with paste of either sodium percarbonate and hot water or liquid hydrogen peroxide and baking soda. First clean the surface with the 3% or general purpose sodium percarbonate cleaning mixture to remove loose debris, then rub in the paste and let it stand for up to an hour, or even a couple of hours, depending how severe the staining is. Rinse and wipe clean with a damp cloth or sponge and repeat if necessary.

An old toothbrush is a handy tool for a good scrubbing of the grout and around the caulked edges of tubs and showers. Be sure to repair missing, cracked or old caulk to avoid allowing water to flow behind the shower or tub unit and create mold issues in the walls and floors.

41. Shower Curtains & Doors

Regular Strength

To remove dirt, grime, mold and mildew from shower curtains start by soaking the shower curtain in a warm to hot solution of sodium percarbonate at regular strength. Let soak at least one hour, then scrub gently to remove the debris.

In cases where the mildew or mold staining is very severe, you may wish to apply the sodium percarbonate in a paste to the stained areas and let set overnight. After treating the stains wash the shower curtain on the gentle or normal cycle of your washer, and then hang to dry.

Shower doors can be sprayed down and wiped clean with liquid 3% solution or the general purpose cleaning strength of sodium percarbonate. For serious soap scum removal add baking soda and scrub the baking soda/peroxide paste over the entire door surface and rinse clean with the shower. (This is why hand held shower heads are amazing for cleanup, because you can point them right where you want them.) Wipe the doors dry after rinsing and make sure you remove all the residue at the base of the door when the job is complete.

42. Toilets

Regular Strength or Double Strength

Cleaning the toilet is another one of those modern chores no one likes and no one wants to do. But someone has to do it! With hydrogen peroxide you can get the toilet bowl clean without introducing toxic chemicals, caustics or acids into your water system. Toilet bowl cleaners are generally among the most toxic of the household cleaners, so it's definitely worthwhile to replace them with hydrogen peroxide.

Start with a clean brush, a spray bottle of 3% solution of hydrogen peroxide, and rubber gloves. Spray down the brush until it is thoroughly saturated, then use it up under the rim of the bowl to scrub away stains. Spray down the rest of the toilet, including adding some more spray to the bowl and let stand for 10 minutes. Return, scrub with the toilet brush, and flush. You can re-spray the toilet with hydrogen peroxide solution after you are done, then allow to dry. This will kill all the surface bacteria and germs, which cause odors and leave the toilet area smelling fresh and clean.

I like to keep a spray bottle of 3% hydrogen peroxide under the bathroom sink so I can give the toilet a good spray-down anytime. This helps to keep the bathroom smelling clean and fresh and also kills all the bacteria on contact. No need to wipe dry, just spray, brush if needed, and leave. Letting the spray air dry will allow the peroxide to get the full time it needs to completely disinfect the toilet too, and no extra work for you to accomplish it.

For more serious staining, spraying 3 percent hydrogen peroxide directly on stained toilet surfaces or applying a paste of sodium percarbonate are both good options. The 3% will not work as quickly, and if the toilet has not been properly cleaned in a long time, the sodium percarbonate may be a better choice because it is much stronger concentration, and even if the toilet bowl is not white, the material it is made of will not bleach, so no worries there.

If using the 3% spray be sure to let it stand for at least 30 minutes before wiping or rinsing.

Regular strength sodium percarbonate solution for cleaning can replace most of the toxins used for toilet bowl cleaning on the market today at a fraction of the cost. For severe staining make a paste of the powder with hot water and scrub the stained surface with it directly, then leave a layer of the paste over the stained area for an hour, rinse and wipe clean.

> ➡ Tip: Spray the toilet brush with 3% hydrogen peroxide before returning it to its holder to ensure the brush is disinfected between uses.

43. Mold and Mildew Killer

Regular Strength or Double Strength

Mold and mildew buildup on ceilings and around bathrooms can easily be removed with a spray of 3% hydrogen peroxide solution. You may choose to allow the peroxide to work for you and come back later to re-spray and wipe clean, or for mild cases of mold and mildew simply spray, wait a few seconds, and wipe clean. Safe for use on walls, floors, tile, and grout (you may wish to spot test wallpaper if in doubt.)

Some molds (particularly black molds) are extremely toxic. For complete instructions on proper and safe removal of black mold see 5. Black Mold Removal under General Cleaning.

Do NOT brush, vacuum or otherwise attempt to remove dry mold, as this will release the spores into the air and spread the mold to new areas of the home.

Laundry

44. Clothing – Spot Remover

Regular Strength

If you've ever spilled red wine, spaghetti sauce, coffee or other highly staining material on your clothing you know that feeling of frustration as you think you've just ruined a favorite shirt, dress, slacks, etc.

In most cases, particularly on fine or delicate fabrics, it seems pretty hopeless. Not anymore. Hydrogen peroxide in a 3 percent solution comes to the rescue. So far, I have not met a stain that I cannot remove using a simple method of spraying the area with 3% hydrogen peroxide and letting it sit; sometimes needing to reapply several times over the course of a day. Wine, red sauce, coffee, chocolate, ketchup, salsa, even mustard stains treated this way disappear without damaging the clothing.

This solution works on any fabric which can be washed with water. When in doubt spot test for color fastness in an inconspicuous area (an interior seam works well).

Another way to apply the peroxide, especially to small spot stains is to use a Q-tip dipped in 3% solution and gently press it onto the stain or spot. This allows more control and uses less peroxide so there is no waste when treating small spots or stains.

In some cases you may want to use a tiny bit of dish or laundry soap at the same time, but most often I simply use straight 3% solution of hydrogen peroxide and it does the job fine by itself.

➡ Tip: NEVER use bar soap to try to remove a stain. Bar soap contains ingredients which act as binders, and you will simply set the stain permanently if you use bar soap, even on a fresh stain. Instead, hang the garment so that the stain is easily accessible and saturate the stained area with 3% hydrogen peroxide solution. Check for results in an hour or so, if the stain is still visible, repeat the saturation (a spray application is easiest).

In one case I had a bright white shirt, which I spilled major spaghetti sauce on, and I really thought that there was no way even hydrogen peroxide could do the trick. It took several applications over two days but in the end the stain was completely gone and the shirt was saved.

Do NOT try to "speed up the process" by using stronger solutions of hydrogen peroxide, as they will bleach the fabric. Instead, be patient and reapply the 3% solution over a day or so. It WILL work!

General Laundry – Sodium Percarbonate

Benefits of using sodium percarbonate in the laundry:

- No environmental hazards - breaks down to oxygen, water and sodium carbonate (soda ash) in your wash water.
- Color safe and fabric safe. It brightens colors.
- Continual use will not cause yellowing or graying of cotton fabric
- Effective stain removal in a broad range of water temperatures
- Does not weaken the strength of fabrics like chlorine bleach
- It is very effective as a laundry presoak for heavily stained articles

In the laundry Sodium Percarbonate is used to remove stains, deodorize, and whiten.

Laundry Applications

It is very effective as a laundry presoak for heavily stained articles. For light soils add 2 ounces of Sodium Percarbonate with your laundry detergent per load. For moderate soils use 3 ounces and for heavily soiled articles use 4 ounces. It is recommended that the sodium percarbonate be added to the washer as the water goes in, before adding the items to be laundered. For front loading machines where this is not possible, mix the sodium percarbonate into one to two cups of hot water and add through the detergent compartment once the washer has started and the laundry soap and water have already run through the compartment and into the machine.

45. Brighten Colors, Color Safe Bleach

Regular Strength

Add one ounce (1/8 cup) sodium percarbonate to your regular wash for brighter colors and whiter whites in mixed color articles.

46. Whiten Old Linens & Curtains

Double Strength

To whiten old linens and yellowed window shears, or to deodorize and remove perspiration stains on garments: soak in 2 ounces (1/4 cup) of Sodium Percarbonate per gallon of warm to hot water for one hour. Swish or stir the mixture with a plastic or wooden utensil a few times during the soaking to make sure all surfaces of the fabric are well saturated. You may continue to soak for up to five hours, after which time the sodium percarbonate has depleted its oxidizing power.

47. Heavily Soiled Items

Extra Strength

For extremely heavily soiled articles, oil soaked rags, work clothes or other deep set grime, mix a presoak in up to 3 ounces (~1/3 cup) sodium percarbonate per gallon of hot water. Allow articles to soak for at least one hour, or overnight. The sodium percarbonate will be "spent" in approximately 5 hours. After soaking, launder as normal.

48. Presoak for Stain Removal

Double Strength

Remove stains from linens, clothing, or bedding using a presoak of 1 to 3 ounces sodium percarbonate in warm to hot water. Let soak for not less than one hour for best results. Launder as usual after presoak.

49. Fine Garments

Regular Strength

To wash or remove sweat and other organic stains from fine fabrics, use a solution of 1 ounce sodium percarbonate per gallon of lukewarm water for hand washing, or simply add approximately 1 ounce (1/8 cup) dry powder to the washer on a gentle cycle setting.

50. Reviving Old Tennis Shoes

Double Strength

Revive old tennis shoes with a presoak in 2 ounces of sodium percarbonate to one gallon hot water. Let soak overnight or at least 5 to 6 hours for best results. Drain and wash in regular cycle in washing machine, air dry outdoors in direct sunlight or dry on low heat in an automatic dryer.

This works particularly well for white tennis shoes and brings them back to a bright white and fresh look and smell. For extremely deep or well set stains, use a paste of sodium percarbonate and hot water, scrub the area and then rinse, or allow to stand for up to an hour before rinsing.

Remember to always start with the lesser time and the lowest concentration of the suggested range and increase time before concentration to get the best results.

51. Most Organic Stains

Regular Strength

Hydrogen peroxide, either in liquid or dry form, is most effective on any kind of organic stain. This includes foods, dirt, oils, most plants, urine, sweat, molds, and mildew, wine, coffee, mustard, tomato, grass stains and blood.

Always start with a regular strength solution (3% liquid or 1 ounce (1/8 cup)) sodium percarbonate to one gallon of water and repeat treatment a second time before using a stronger solution. Allowing solutions to soak for one hour or longer may be all that is needed to remove most organic stains.

Using liquid hydrogen peroxide at room temperature or sodium percarbonate solutions at warm to hot temperatures will provide the best results.

Sick Room

52. Hand Wash

Regular Strength

Replace "anti-bacterial" soaps, chlorine based wipes, sprays, disinfectants and other toxin-laden products with simple soap and a spritz of 3% hydrogen peroxide solution.

Just wash your hands with natural hand soap and then spray them with the hydrogen peroxide and wipe dry.

You can also add peroxide to a liquid hand cleaner, and save the second step. Just wash hands as normal with the liquid soap and peroxide mixture.

53. Spray Disinfectant

Regular Strength

Use 3% hydrogen peroxide solution as a spray disinfectant around the sick room, to clean surfaces, and to stop the spread of infection.

When dealing with seriously contagious environments use the disinfection regimen as described in the kitchen section of this book, one spray bottle each of hydrogen peroxide and white vinegar. Spray one after the other (the order does not matter) and wipe dry or let air dry. This will kill 99.9% of all pathogens, bacteria and microbes including staph, strep and E-coli.

See the personal care section for effective cold prevention with hydrogen peroxide for added protection when taking care of family members with colds and flu.

54. Sore Throats

Regular Strength

Use a 3% solution of hydrogen peroxide as a mouthwash and gargle when you have a sore throat. It will kill the infection on site and speed healing. If you find that 3% is uncomfortable (your throat is really raw and sore) dilute the mixture by 50 percent or more with distilled water until you find a comfortable dilution for yourself. The peroxide will still work.

55. Vaporizers

Regular Strength

Adding 3% hydrogen peroxide to the vaporizer is an excellent way to put additional oxygen into a sick room, and improve ease of breathing during congestive illness. Simply add one cup of 3% hydrogen peroxide when filling the water chamber to the standard fill line.

Between uses, the spray bottle of 3% solution is the perfect healthy and natural cleaner for the vaporizer before it is put away.

Do not heat or bring hydrogen peroxide to a boil. Adding a dilution of 3% to a vaporizer reservoir is fine, but do not use even 3% peroxide in a straight undiluted form in the vaporizer.

Baby Care

56. High Chair Trays

Regular Strength

Keep a spray bottle of 3% hydrogen peroxide handy for quick cleanup and disinfection of baby's high chair tray. The hydrogen peroxide is safe, non-toxic and effective at killing germs, so baby's eating surfaces will be clean without exposure to dangerous chemicals.

If you need a little viscosity in the mix for stuck on foods, add a few drops of natural dish washing liquid to the spray bottle and shake gently.

57. Playpens/Cribs

Regular Strength

Wipe rails, flat surfaces and interiors with a 3% hydrogen peroxide solution and a clean cloth. Because babies often teethe on the rail edges of their cribs or playpens and always are putting their hands in their mouths even if they aren't teething and chewing the actual crib or playpen, it is important to keep these surfaces clean. A quick wipe down every few days with a 3% solution left to air dry will disinfect and clean at the same time.

Pre-soak baby's bedding in a 3% liquid solution or in a regular strength sodium percarbonate solution to disinfect and eliminate odors. For urine soaked bedding, presoak in a 3% or regular cleaning strength sodium percarbonate solution with equal parts baking soda for complete odor elimination, then launder as usual.

58. Diaper Pails

Double Strength

Whether or not you are diapering your baby in cloth diapers, there is no question that cloth diapers are very handy for general baby care.

If you are using cloth diapers to diaper your baby, the chances are that you were taught to use chlorine bleach to wash them. While the disinfection certainly is handled this way, the addition of chlorine to cloth which will be worn next to baby's skin is a less than optimal result. Chlorine is absorbed by the skin, and babies are particularly sensitive and susceptible to skin irritation from harsh chemicals.

To replace chlorine bleach in diaper pails use 1 cup of 6 percent hydrogen peroxide liquid or 3 ounces sodium percarbonate for every gallon of soaking solution for the diaper pail. The solution should also contain simple laundry soap – Original Arm and Hammer Washing Soda (also known as Soda Ash, or Sal Soda in some places), Dreft, or some other mild laundry detergent formulated for baby's skin.

For soaking pails, change or add fresh solution daily to retain antibacterial properties of the hydrogen peroxide. Diapers should be well rinsed of all fecal matter and debris before placing in diaper pail.

Add 2 ounces (1/4 cup) sodium percarbonate to the washer with hot water and laundry soap before adding the diapers to the washer for normal laundering.

59. Baby Laundry

Regular Strength
or Double Strength

Add 1 ounce (1/8 cup) of sodium percarbonate to regular washing to get baby's clothes fresh and bright and remove odors of sour milk and/or urine.

For heavily soiled baby items, presoak in hot water (100 to 150 °F) and sodium percarbonate at a dilution of 2 ounces per gallon of water. Let soak at least 30 minutes before laundering.

60. Baby Bottles

Regular Strength

To disinfect baby bottles after washing in hot soapy water spray the interior of the baby bottles with 3% solution of hydrogen peroxide. Set bottles upside down in dish drainer to dry. No rinsing is required.

This same technique can be used for bottle nipples and pacifiers, or these smaller items can be set in a rinse water of 3% hydrogen peroxide for 5 to 15 minutes after washing, then simply remove, drain and set to air dry.

61. Toys

Regular Strength

To keep baby's plastic toys, blocks and playthings clean, spray with 3% solution and wipe dry with a clean cloth. It is also a good idea periodically to soak all plastic toys in a regular cleaning solution of sodium percarbonate for one hour. Water should be hot, and baking soda can be added to remove all odors from the plastic.

62. Pacifiers/Teething Rings

Regular Strength

Items often in baby's mouth, like teething rings, should be given a 3 percent hydrogen peroxide bath occasionally to kill germs and sanitize thoroughly. The easiest way to do this is to fill a small to medium sized Tupperware container with enough hydrogen peroxide to cover the teething ring or other teething toys.

Wash the items to be sterilized with hot soapy water and rinse, then simply place in the hydrogen peroxide "bath" for a minimum of 20 minutes. Remove and allow to air dry.

Personal Care

63. First Aid

Regular Strength

Hydrogen peroxide is most commonly known for treating wounds, open cuts and sores, or infections on the skin. It works great for these purposes, and is highly recommended. At 3% solution it does not act instantly and should be allowed to bubble up on the surface to do its work.

For most common household cuts and scrapes, hydrogen peroxide is all you'll ever need... But in cases where known pathogens were present at the wound site (coral reef cuts, for example) it is wise to apply an antibacterial antibiotic ointment such as Neosporin or erythromycin to the wound after cleaning.

64. Bathing

Regular Strength

To rejuvenate stressed nerves, tired muscles and sore aching joints, there is nothing much better than a good soak in the tub with hydrogen peroxide and either Epsom, sea, or mineral salts.

Hydrogen peroxide baths have been a staple for many in the naturopathic community for some time. In fact, it is now known that the healing waters of Lourdes, Fatima's and other famous "healing springs" around the world have higher than average levels of hydrogen peroxide.

So soak your worries and aches away in a hydrogen peroxide bath and you'll feel better, soften and heal your skin, and sleep like a baby.

A standard hydrogen peroxide bath can be made of 1 to 2 cups of 3% hydrogen peroxide to approximately one half tub of water in a standard bath tub size. If you have an oversized tub or "garden tub", you'll just need to add more hydrogen peroxide. Add one cup of either Epsom salts, mineral or sea salts and you'll give yourself a really healthy mineral soak.

When my youngest son used to get migraine headaches, I would draw him an Epsom salt and hydrogen peroxide bath and turn the lights down low in the bathroom so he could lie in semi darkness in the tub. It always reduced the headache, and if he would stay long enough, adding more warm water over time (an hour or so) the headache would really disappear and I'd tuck him into bed with an herb tea to get a good sleep. This almost always worked well enough that he awoke without a headache after a good sleep.

Of course the challenge with children, even when they have migraines, is to get them to go along with the remedy – but once he learned that it worked, that was not a problem.

This bath is great not only for soothing the nervous system, it also works wonders for your skin, healing small abrasions and oxygenating the whole body. Some even say that regular hydrogen peroxide bathing will eliminate parasites in the intestinal tract and throughout the body. While I have never done any research on this aspect of hydrogen peroxide bathing, I can say that as a regular user of hydrogen peroxide for bathing, ear drops and mouth wash, I have not come down with any of the colds or flu's other members of my household have been stricken with over the years. I just don't get sick, or if I get a slight cold while caring for others who are in bed for a week, it lasts a day and is gone again. Well worth the "effort" of using hydrogen peroxide in my self-care regimen.

Another important aspect of bathing with hydrogen peroxide and mineral salts is that it provides an excellent way to deliver minerals to the body. Whether done through full body soaking or foot soaks, the evidence is clear that the addition of magnesium salts, for instance, can greatly reduce magnesium deficiencies, which, according to the World Health Organization, affect approximately 80% of the world's population. See more about mineral salts in the General Foot Care section (Item # 71).

65. Teeth

Regular Strength

No, we are not talking only about bleaching teeth. Bleaching teeth is routinely accomplished with hydrogen peroxide and other forms of peroxides which are commercially available in strips, gels, or at the local dentist in specialized applications.

If bleaching your teeth is what you want to do, it is HIGHLY recommended that you stick to commercially available applications designed for this purpose. Do not attempt to make up your own teeth bleaching solutions unless you have a great deal of experience using hydrogen peroxide; and even then I would simply avoid any risks of home created remedies and use a commercially available product designed for safe teeth bleaching.

Hydrogen peroxide at levels above 5% can cause bleaching in fabrics and on skin. That means the 3% solution you may use as a mouthwash or dental orifice is not likely to actually bleach your teeth, however, regular use of good old 3% solution in teeth brushing and mouth rinse can be very effective at keeping teeth bright and clean. If you need more whitening power than this, please remember that teeth are made from enamel; they are not impervious to oxidation which can, conceivably, at higher concentrations, actually damage or wear away the enamel itself.

No matter how tempting, do NOT ever use a straight hydrogen peroxide solution higher than 3% solution directly on the teeth surfaces. Stick to the tested and safe products on the market designed for teeth bleaching to do that job.

Do use the 3 percent solution in a mix with baking soda for toothpaste and you'll discover your gums get healthier, your breath is fresh and your mouth is all around healthy.

66. Mouthwash

Regular Strength

It is generally recommended that hydrogen peroxide solutions be used to heal cuts and sores in the mouth by gargling with a dilution of 3% hydrogen peroxide in water after teeth brushing.

There are those who suggest that one can gargle with straight 3% hydrogen peroxide, but we have never found this to be comfortable, and so have always used a roughly 1.5% solution for our home mouthwash.

Here is the recipe for the home mouthwash we make and use: Mix 32 ounces filtered or distilled water, 1.5 ounces (three tablespoons) of 35% food grade hydrogen peroxide and 8 drops of peppermint essential oil.

Perhaps you will want to use a little less peroxide, or eliminate the peppermint oil. It's safe and easy to experiment so long as you never forget the basic rule that nothing over 3% solution should ever go on or in your body.

If you are using a solution of food grade hydrogen peroxide that is lower than 35% to make your mouthwash use the conversion table in Appendix C to determine how much peroxide at the percentage solution you are using is needed to make a 1.5% mouthwash.

67. Hair Bleaching

Double Strength

Many people mistakenly believe that they can achieve a "platinum blond" hair color by using straight hydrogen peroxide (at 6% to 9%solution) for bleaching. In fact, while hydrogen peroxide will strip the hair of color, or lighten it, it does not make the hair blonde so much as "no color" which looks more or less like a cross between white and grey and does not look really "blond".

To achieve those rich blond tones you need actual hair coloring agents. Hair coloring agents and hydrogen peroxide (mixed with other activating bleaching chemicals) is what most commercial hair color products are made of. They are full of chemicals, and pretty darn toxic, but hey, if the color of your hair matters that much, go ahead and use a commercial product to bleach your hair.

Or, if what you really want is just hair without color, you can "bleach" your hair with hydrogen peroxide. First snip a small lock of hair for testing. Then mix a solution of 6% to 8% hydrogen peroxide, and dip the test lock of hair into the solution. Place the wet strand of hair on a clean surface and wait approximately 15 minutes to ½ hour while observing the change in hair color of the strand. Do not use a solution stronger than about 9% for this purpose, and be aware that applying this solution to your hair will definitely mean it will be in contact with your scalp and for most people the sensation is one of slight tingling to actual "burning". The hydrogen peroxide will not do any permanent damage to your scalp or your brain or anything like that, but it doesn't feel so great for some folks!

In general, you will observe a lightening of the hair. However, depending on the original color of the hair and other previous chemical applications, etc. the change in color may not be what you expected.

For those with naturally deep red hair color, a slight yellowing or bronzing effect may occur. For those with brown or darker hair a sort of transparent "non-color" may occur. Test that strand before you attempt to change the color of your whole head of hair to avoid unhappy surprises!

68. Contact Lenses

Regular Strength

Hydrogen peroxide solutions have been available for contact lens disinfection since the late 1990's - early 2000's. There are various products available and all of them are required to be tested to comply with a strict set of ISO standards based on performance, purity and quality of ingredients.

As a long term contact lens wearer, there is no question that the hydrogen peroxide based cleaning system provides clearer, cleaner and much sharper focus of the lens. No doubt the bubbling action of the solution removes more of the protein and other filmy deposits typical of contact lenses over time.

There are several levels of efficacy required to properly clean and disinfect contact lenses for safe wearing practices. These are designed to ensure that disinfection adequately destroys pathogenic microbial, bacterial and fungal organisms which can cause damage to our eyes.

The most efficient and effective peroxide-based defense against organisms such as Acanthamoeba polyphaga (which can cause serious eye damage or even blindness) requires a two-step disinfectant and neutralization process.

Hydrogen peroxide is an extremely effective anti-bacterial, microbial and fungal agent. However, it cannot be placed directly in the eye without damaging effects.

Therefore, in order to utilize its substantial efficiency to protect your eyes as a contact lens wearer, the peroxide must be neutralized on the lens before the lens is inserted back into the eye.

The modern solution for this is a neutralizing tablet or component of the actual lens case which begins the neutralizing process as the disinfecting process begins. While this process is perfectly effective in normal situations, it is less effective than a direct hydrogen peroxide exposure first, and neutralization after disinfection.

Additionally, lenses stored in a neutralizing hydrogen peroxide case over time have no protection against infectious bacteria or microbes as the neutralizing element of the lens cases (platinum) has effectively neutralized the peroxide within the first 6 hours of storage and no disinfecting properties remain in the solution in the storage case after this time.

Perhaps the manufacturers should develop a non-neutralizing case for use for storage purposes beyond daily wear, so as to accommodate those lens wearers who do not wear their lenses every day and therefore lose the benefits of the disinfecting power of the peroxide to the neutralizing agent before long term storage begins.

For contact lens wearers who use the hydrogen peroxide based disinfection system which employs neutralizing tablets which are added to the solution in the case, it would be possible to delay the addition of the tablet when lenses are being stored for longer periods, to ensure proper disinfection continues during long term storage. However, then it will be a necessity to neutralize the solution with a tablet no less than six hours before intended use of the lenses.

Regardless of the method chosen, the neutralizing while disinfecting system does still work perfectly adequately for short term contact lens storage; however, it is less effective than a separate and distinct disinfecting period prior to beginning the neutralizing process. Unfortunately, products which were designed to provide these steps separately were not popular with lens wearers who opted for convenience rather than safety in their contact lens care protocols. These products are therefore no longer available.

For this reason, it is generally suggested that long term storage of contact lenses utilize the chemical multi-purpose solution rather than hydrogen peroxide based disinfection systems.

To read further about contact lens solution comparisons see "Comparison of Hydrogen Peroxide Contact Lens Disinfection Systems and Solutions against Acanthamoeba polyphaga" by Reanne Hughes and Simon Kilvington (http://foodgradeh2o2.com/hydrogen-peroxide-contact-lens-disinfection.)

69. Cold/Flu Prevention Ear Drops

Regular Strength

This is one of my longest standing and all-time favorite hydrogen peroxide applications and it works like a charm, if you do it soon enough. If you wait until you are already really sick it will work, and shorten the time of your illness, but it won't prevent it, and it won't stop it in its tracks exactly... You'll still need a day or two to get better.

So before you really are sick, if you

- Notice others around you in your household are sick
- Notice that your throat is starting to feel "scratchy"
- Notice that your nose is getting stuffy
- Notice that your sinuses are feeling enflamed or sensitive
- Notice that your eyes are watering or you're sneezing

At the first sign of any of these "cold precursors" it's time to take ten minutes and stop the cold or flu before it really gets hold of you.

Here is what you'll need:

- Several Q-tips and a small dish, or a small eye dropper and bottle
- A handful of tissues

Lie on one side and drop six to eight drops of 3% hydrogen peroxide solution into your ear (the raised ear, obviously.) Lie on your side for at least 3 to 5 minutes and let the hydrogen peroxide trickle down inside your ear canal. This feels really funny at first and for some people the tickling is just too much, but TRY to get used to it, and let that bubbling and gurgling go on for at least a few minutes and up to five minutes per ear. You may or may not hear the bubbling, depending on how stuffed up you are and how much peroxide is in the ear, but you will definitely feel it. The further (deeper) into the ear canal it goes the better it seems to work; just be sure to take adequate time to let the ear drain by turning the head the other way after each application.

To drain the peroxide out of the ear, apply a soft wad of tissue to the ear as you turn your head to let the hydrogen peroxide drain out. It will all drain out over a few minutes quite readily, while you apply the hydrogen peroxide drops to the other ear. Then turn again and repeat the draining process to allow the second ear to drain.

Do not simply stand up and tip your head to drain the second ear. Really stay lying down and turn your entire body and head and wait 3 to 5 minutes to allow the second ear to drain.

When it's flu season and the kids are bringing home colds and coughs and flu from school it's a good idea to repeat this remedy every few days or so to prevent catching everything the kids are bringing home.

It's a simple process and well worth the few minutes it takes to avoid spending a few days to a week in bed just because there are colds going around.

70. Hand Washing

Regular Strength

Mix in 1 part hydrogen peroxide to 3 parts of your favorite natural liquid hand soap, shake or mix well for added disinfection power. This will eliminate the 'need' for anti-bacterial soaps, which should be avoided.

Be sure to read *"The Newest Line of Antibacterial Soaps and Why We May Want to Avoid Them"* in the earlier chapter *"From Toxins to Oxygen"* regarding anti-bacterial soaps and their impact on our environment and particularly in the soils where our foods are produced.

71. General Foot Care

Regular Strength

Just as it does wonders to soak your whole body in a hydrogen peroxide bath, so it does your feet great wonders to soak them in a tub of hot water, hydrogen peroxide and Epsom or mineral salts.

If you suffer from sore, tired feet, aching calves, or knee pain, the addition of magnesium oil (magnesium chloride solution) to the soak can provide excellent and quick pain relief too.

We've been using these hydrogen peroxide and mineral salts baths for decades now and they make all the difference in the world, especially after really physical or emotionally stressful work. Even a thirty minute soak can leave you so rejuvenated you will sleep deeply and well and wake up feeling refreshed.

In fact, according to recent reports from the World Health Organization and other researchers, magnesium deficiency affects approximately 80% of all adults worldwide. It is so serious, in fact, that in Finland a national program was launched to substitute magnesium salts into the national diet which resulted in Finland's ranking as number one for congestive heart failure to drop to the number 7 position.

Clearly, magnesium is critically important to our health, and most of us are not getting enough of it. Interestingly, statistics show that this is a relatively new problem, the result of our move away from traditional food and farming practices to mass produced food products. Whatever the cause, it is clearly worth our time to get more magnesium into our systems, and topical sprays or hydrogen peroxide and magnesium salt soaks are great ways to accomplish this.

Learn more about magnesium oil, magnesium deficiency and the advantages of topical magnesium applications at "Ancient Minerals" (http://foodgradeh2o2.com/ancient-minerals-magnesium-deficiency) and "Magnesium for Life" (http://foodgradeh2o2.com/dr-sircus-magnesium-articles)

You can either add the magnesium salts to the soak water or use a topical magnesium spray to spray and massage in the magnesium oil before soaking your feet in the hydrogen peroxide foot bath.

Magnesium oil is really a liquid salt solution of magnesium as it occurs in natural underground lakes. It does wonders for the body. Soaking your feet in a magnesium salts and hydrogen peroxide warm water foot bath is a great way to get more magnesium into your system. Magnesium is a natural pain killer in the body, helping to eliminate joint pain, muscle pain and bruises.

Additionally, doctors have now proven that regular foot soaking in hydrogen peroxide actually improves circulation to the feet, reducing varicose veins and swelling and killing the fungi and other bacteria associated with athlete's foot.

A good foot soak in a whirlpool foot bathing tub with hydrogen peroxide and minerals feels wonderful and has a very calming and soothing effect, too.

After feet are dry, spray with a magnesium oil/salt solution to improve magnesium levels, increase joint flexibility, reduce bruising and relieve muscle pain.

And speaking of athlete's foot…

72. Athlete's Foot

Regular Strength

Before you go out and spend a small fortune on who knows what chemicals, drugs or other remedies for athlete's foot, try soaking your feet as noted above.

Spraying the feet with straight 3% hydrogen peroxide solution after bathing and at the beginning and end of the day after wearing shoes or after showering, will dramatically improve all forms of scaly skin, athletes foot and other foot problems.

Spraying down the shower stall you use daily after each use with 3% hydrogen peroxide solution will also kill the fungi that cause the problem and help to eliminate repetitive re-infection.

After soaking the feet, use either a pumice stone or a drug store variety skin file to remove the excess dead skin that will become soft and white during the soaking process. Then scrub feet briskly with a clean foot brush or bathing sponge and hydrogen peroxide before drying.

73. Skin Infections and Rash

Regular Strength

A hydrogen peroxide bath is the best cure for skin abrasions I know, and I've had scrapes heal up overnight after one hydrogen peroxide bath, but if you don't have the time or inclination to soak in the tub, just spritz the affected area a couple of times a day and let air dry.

Your skin will heal much faster, and as a benefit I've noticed there is no scarring.

74. Circulation Improvement

Regular Strength

This remedy for circulation improvement is from Majid Ali, M.D.

If you are interested in understanding the complexities and intricacies of integrative medicine, his works are a great place to start. We've put his web site information in the Appendix of Resources at the back of this book.

Protocol for Hydrogen Peroxide Foot Soaks and Baths

Hydrogen peroxide soaks can be used with different concentrations of H_2O_2 and salt. The following is the standard protocol prescribed at the Dr. Ali's Institute:

H_2O_2 Soaks Protocol

- Water: 20 parts
- 3% H_2O_2: 1 part
- Salt: one teaspoon
- Time: 20 minutes

The recommended choices of salt are as follows: (1) Epsom salt; (2) sea salt; and (3) common table salt.

Stronger solutions of H_2O_2, such as one part of H_2O_2 and 10 parts of water or 1 part of H_2O_2 and 15 parts of water may also be tried to test for variations in efficacy for individual persons.

For chronic conditions, I generally prescribe foot soaks on a four or five day a week basis. For sub-acute conditions, daily soaks are recommended. Uncommonly, I have prescribed such soaks on a twice-daily basis.

There are several good brands of foot-soak and foot-massage units available on the market. The one made by Brookstone Company creates effective whirlpool conditions and includes a "nodule" for effective massaging of tender points on the feet or ankles.

75. Skin Care – Anti-Acne

Regular Strength

Acne breakouts – everyone hates them, but lots of us get them. Even occasional breakouts are a nuisance.

Real acne is minor clogging of pores with sweat and dirt. It is not infectious, and is not a skin eruption from below the dermis. A thorough washing of the face with a mild cleanser each day is the best deterrent to acne.

For spot relief, a cotton swab dipped in 3 percent hydrogen peroxide can disinfect a pimple… Do not apply pressure. If the hydrogen peroxide has no effect, then the eruption is something more systemic than acne and may require other treatments.

In fact, if you have something you are thinking of as "acne" and it does not respond to hydrogen peroxide application then you very likely are not experiencing an acne break out. It is more likely to be a form of boil, infection or other skin irritation, which is symptomatic of the body's release of toxicity and not simple acne.

This is good to know if you've been thinking to yourself that hydrogen peroxide doesn't "work on my acne".

Hydrogen peroxide heals acne with only one or two applications to the affected area.

For general good skin care, it is also great to spritz with 3 percent hydrogen peroxide after showering or bathing.

If you have very sensitive skin, dilute your 3 percent solution again by adding 50 percent distilled water, and it will still be very healthful for your skin.

76. Deodorant

Regular Strength

Because hydrogen peroxide is anti-bacterial, it is a great deodorant. It simply kills the germs and bacteria that cause odors.

A hydrogen peroxide spray or quick wipe with a hydrogen peroxide saturated cloth allowed to air dry and then followed by crystal deodorant stone or other natural deodorant will keep you smelling fresher longer, since you've started with a clean skin surface.

77. Bruise Soak

Regular Strength

The next time you bang your shin, or whack your wrist, set it to soak in a soak of mineral salts and hydrogen peroxide mixed in hot water. It is possible to completely "skip" the bruising stage this way, if done right away.

Mix 2 to 4 tablespoons magnesium oil or Epsom salts, 4 ounces 3% hydrogen peroxide in one to two gallons hot water. If you start with one gallon and keep adding water, you may get to two gallons… see?

This soak is also great for circulation, and the recommended foot soak protocol from Dr. Majid Ali can be found there.

Plants

78. House Plants

Regular Strength

Keep houseplants healthy and green by adding 4 ounces (½ cup) 3% hydrogen peroxide to 1 gallon of water when watering. This will oxygenate the soil and root system of the plants, improving growth and color.

If you have a plant that is not doing well, repot it in fresh soil and use this watering solution intermittently for several weeks.

79. Plant Diseases & Fungi

Regular Strength

Use hydrogen peroxide as a natural fungicide to kill diseases and microorganisms attacking your houseplants. Mix 8 ounces (1 cup) 3% hydrogen peroxide to 1 gallon of water. Dispense in a spray bottle, misting the affected plants every few days until they improve.

In cases of serious damage or disease, many horticulturists recommend a straight 3% spray solution applied directly to plant leaves. However, some plants are very tender, and this solution may be too strong. Test on one leaf before spraying the entire plant if you are uncertain.

The 8 ounces 3% to one gallon of water solution is also recommended for plants in high stress from either too little water, or over watering leading to root rot. In the case of over watered or improperly drained plants, repot in fresh soil before applying the solution.

80. Sprouting Seeds

Regular Strength

Just as hydrogen peroxide helps with growing tasty sprouts for eating, it is also a great helper when it comes to sprouting seeds for planting.

Whether you are getting ready to sow seeds into early spring indoor flats or directly out into the garden, a soak in a mix of water and 3% hydrogen peroxide mixed in equal parts can give seeds the boost they need to get moving.

To start larger seeds (corn, squash, beets, spinach, peas, beans etc.), you can lay out the seed on a doubled paper towel and then fold over the towel on top of the seed and saturate the paper towel. Place it on a plate or shallow dish and keep moist for a few days until you see the little spurs emerge, then plant as normal.

To start smaller seeds (peppers, chards, brassicas, celery, lettuce, carrots etc.), use equal parts peroxide and filtered water in a shallow dish, place the seeds in the dish and cover completely. These can soak for four to six hours or overnight, then plant as usual.

Using this method with celery seed, which can be very slow to germinate, particularly in cooler weather, can drastically speed up sprouting time, and this is also true of peppers which can be slow to get started.

Outdoor Uses

Garden

81. Garden Plants

Regular Strength

Hydrogen peroxide works equally well in the garden, and can be applied using a watering can or with a liquid sprayer. Set the dilution to 4:64 for liquid sprayer units (that translates to 1 ounce per gallon, or roughly 1/8 cup per gallon.) Saturating the soil around garden plants and spraying the foliage will improve plant strength, growth and color.

Do not increase the strength of the solution when spraying hydrogen peroxide in the garden, as it will upset the balance of the living organisms in the soil and can kill off good microbes needed in a healthy garden soil. At the lower concentration of 1/8 cup per gallon it will not disrupt the natural healthy soil microbial processes but will oxygenate the soil.

82. Cuttings and Rootings

Regular Strength

Start new cuttings to root in a solution of ½ cup 3% hydrogen peroxide to 1 gallon of water. They'll root faster and begin new leaf growth sooner, too.

This also works well for cut flowers or branch clippings from flowering shrubs like lilac or Daphne odora for the table. Use the same ½ cup per gallon of water solution for cut flowers and cuttings.

83. Weed Killer

Extra Strength

Hydrogen peroxide at a 10 percent solution is an excellent non-pervasive and eco-friendly "herbicide". There are no commercial products available at a 10 percent solution, so the easiest way to do this is to purchase 30 percent agricultural hydrogen peroxide and simply dilute it down to the desired 10 percent solution. Apply directly to the targeted weeds. This will also work with aquatic weeds. However, if killing off weeds in a contained fishpond, do not add such concentrations to the pond without first removing the fish and any plants you do not intend to kill! You will also need to test the hydrogen peroxide levels in the pond before returning the fish and other plants to the pond.

We have used this method successfully to kill weed sumac trees which can be very invasive and have roots which just re-sprout a new weed tree if you do not remove the entire root. Because the root can be six feet deep in the ground even in very small sumacs, it is nearly impossible to get the entire root out. Instead, clear a small hole around the root and cut it cleanly across the top. Then use a sharp knife and cut an X into the top of the root. You can pour straight 35% hydrogen peroxide solution into the cut, about a half cup to a full cup depending how large the root is, and leave it exposed. The root will die and then you can choose to remove it or just cover it over with soil. This method also works with straight white vinegar if you would prefer to save your food grade hydrogen peroxide for food preparation and cleaning in the kitchen. When using vinegar you will want to use one cup in the first application and then another cup 24 hours later.

84. Birdbaths

Double Strength

Is your birdbath full of algae and slime? If it is... empty the water and let it stand for a few days in dry weather. Then fill to cover the dried growth with 3% solution of hydrogen peroxide and let it stand for at least an hour. In the years I have been doing this, I've never had a problem with birds going to the birdbath while I'm in the midst of the cleaning, but you may cover the birdbath while the hydrogen peroxide is doing its work. (If the birdbath is in direct sunlight, definitely cover it, as the sunlight breaks down the hydrogen peroxide and will reduce its effectiveness. After an hour or more, scrub out the now mostly broken down algae and crud and rinse with fresh clean water. If your birdbath is really bad, you might have to repeat this process once more to get it really clean.

The added benefit to using hydrogen peroxide to clean the birdbath is that hydrogen peroxide is anti-bacterial, anti-viral and anti-fungal which means that by using it in the birdbath you are helping the birds to remain healthy and to prevent the spread of bird flu among the wild bird population.

To go one step further in helping your native birds stay healthy, put a quarter cup (2 ounces) of 3% hydrogen peroxide into their fresh clean bath water each time you fill the birdbath. The ratio should be approximately one ounce of peroxide per quart of water. You can measure how much water the birdbath holds by using a quart container to fill it one time. Most birdbaths hold approximately a half-gallon of water, but measure if you are unsure or if your birdbath is particularly large or small and adjust the peroxide accordingly.

It won't harm the birds, and it will help prevent the spread of pathogens and disease among them.

Pools & Spas

85. Hot Tubs

Double Strength or Extra Strength

There is nothing as relaxing as a warm, bubbling soak in a hot tub... Particularly if that hot tub is cleaned with hydrogen peroxide rather than chlorinated. The oxygen enriched water smells clean and fresh, softening your skin and soothing your body as you soak.

According to the Merck index, hydrogen peroxide can be used as a water disinfectant. In fact, it is used internationally for water disinfection, treatment of wastewater, water gardens and, increasingly, in swimming pools and spas.

Some newer pool disinfection systems actually use recently developed equipment to generate oxidation in the water as it passes through the cleaning system. In these newer systems the need for additional chemicals in the water can be completely or nearly completely eliminated.

While older spa systems rely on harsh toxic chemicals which fill the surrounding area with their fumes and odor, these newer systems provide clean, fresh oxygen enriched water for bathing which has no odor.

For those not ready to invest in an entirely new hot tub filtration and water disinfection system, food grade hydrogen peroxide offers a transitional solution.

You can eliminate the use of chlorine or bromine chemicals in the spa and use hydrogen peroxide instead of these chemicals. Adding any type of ozonator or UV sterilizer to the system will also assist the hydrogen peroxide in the event that your water contains high levels of iron or organics, which will break down the hydrogen peroxide more quickly. If you are unsure of the mineral content of the water, begin using the hydrogen peroxide as described here, and test for hydrogen peroxide levels frequently.

Begin by shocking the tub with a high dose of 35% food grade hydrogen peroxide. Add one cup (8 ounces/250 milliliters) of 35% hydrogen peroxide per every 250 gallons (1000 liters) of water in the tub. Run the pumps to circulate the water as you add the hydrogen peroxide and then intermittently over the next 24 hours.

➡ Note: Be sure to check and empty the filters when beginning and several times throughout the first 24 hours as the hydrogen peroxide will break down organics and other materials in the water and may at first create an excessive load on the filter system as you transition.

Allow the water to stand overnight (after the initial 24 hours have passed.) Then circulate the water briefly before using a hydrogen peroxide test strip to measure the level of hydrogen peroxide in the water.

Hydrogen peroxide levels should run between 30 and 100 ppm (parts per million) for regular hot tub use. If the levels are below 30 ppm when testing, add hydrogen peroxide at a rate of 1 cup 35% food grade hydrogen peroxide per 500 gallons of water. Circulate and let stand several hours before testing after adding hydrogen peroxide.

By testing often in the early stages of using hydrogen peroxide you will be able to determine how often you will need to add hydrogen peroxide to the spa. The levels will vary according to the frequency and number of people using it. Test at least weekly once you have a general idea of what your spa needs to maintain optimum levels of hydrogen peroxide.

The conversion of hot tubs to hydrogen peroxide has been one of the most commonly asked about topics since the first edition of this book was published in 2006. There have been many phone calls, emails and consultations among hot tub users, myself and others to determine causes of various conditions in the tubs and problems with conversions which did not go well. We have discovered, by trial and error and many users experiences that hot tubs which have been running on bromine solutions often develop deposits inside their filtering and circulation piping which cannot be removed and which reacts with the oxygenation of hydrogen peroxide in strange and peculiar ways: turning water orange, creating a foul odor, etc. At this point in time we do not recommend that a bromine hot tub system be converted to hydrogen peroxide, because it is nearly impossible to get to the pipes internal to the system to remove the deposits of bromine.

If you are shopping for a secondhand tub with the idea of using hydrogen peroxide in it, be sure to ask the owners what chemicals they were using in it before you buy it. Unfortunately, many tubs are now designed for or using bromine, which is no less toxic than chlorine according to my own research, but which does mean conversion to hydrogen peroxide can be problematic at best and a complete failure at worst.

86. Swimming Pools

Double Strength or Extra Strength

Swimming pools as well can be run on hydrogen peroxide instead of chlorine, bromine or other chemicals. Using food grade hydrogen peroxide for swimming pools has become so popular that it is the fastest growing market segment of consumer purchases of food grade hydrogen peroxide.

Like hot tubs and spas, swimming pools can be effectively maintained using hydrogen peroxide. However, unlike hot tubs and spas, swimming pools are substantially larger bodies of water and ideally, should not require constant circulation or pumping.

In recent years, several product lines have become available on the market, which allow you to set up oxidizing systems which do not require the addition of any chemicals, or even hydrogen peroxide. These new systems are becoming increasingly popular as more and more commercial and home swimming facilities recognize the drawbacks of chlorination.

To achieve the highest level of purification it is generally recommended that any oxidizing system should also consist of an additional UV (ultra violet) disinfection system as well as the oxidizing unit. Many pre-packaged systems are comprised of both the oxidizing and the ultra violet disinfection systems.

Be sure to read the section on hot tubs regarding bromine before planning for or making a conversion for your pool. I do not have any reports of problems with swimming pool conversions. However, it would be wise to check with any of the companies providing the newer oxidizing systems for pools regarding their compatibility with bromine systems if you are currently using bromine.

87. Water Gardens

Regular Strength

Hydrogen peroxide is used in commercial water gardens around the world for water sanitizing and as an algaecide. When using hydrogen peroxide for water gardens, a dilution of 8 ounces (1 cup) of 3% solution to each 5000 gallons of pond water will give you a clean water garden. Please note that this is a very weak dilution. One cup (8 ounces) 3% hydrogen peroxide solution to five thousand gallons. This dilution is low enough that it is tolerable to most healthy aerobic life forms in the water garden and will only be toxic to those pathogens which are anaerobic (non-oxygen breathing), which is the whole idea. DO NOT dose the water garden directly with undiluted peroxide. Introduce the dilution by adding small amounts of peroxide pre-diluted in several gallons of water at a time to avoid shocking the natural system of the garden.

You can provide a natural hydrogen peroxide 'infusion' system by adding a small bale or bundle of dry barley, peat (peat will require a mesh netting to avoid simply falling apart in the water) or lavender to the pond. These plants naturally produce hydrogen peroxide in the pond over time.

88. Fish Ponds

Regular Strength

As with water gardens, commercial fish farms, aquariums and other commercial aquatic operations rely heavily on food grade hydrogen peroxide for cleaning, sanitizing, and providing algicidal action to the water.

In smaller consumer based applications, again care must be taken to avoid over dosing the pond. Hydrogen peroxide even in very weak dilutions can affect fish and other aquatic life.

In general, dilutions of 1/8 cup of 3% hydrogen peroxide to 1 gallon of water are preferred, again with a target dosage equal to 1 cup of 3% peroxide to 5000 gallons of water. Diluted mixture should be added to the fish pond one gallon at a time; never change out the entire pond and refill with hydrogen peroxide diluted water. Instead, simply add the hydrogen peroxide, one gallon of dilution mix at a time, over the course of a week or two. There is a newly approved product available for use with aqua culture operations which may be helpful for those with fish diseases in fish ponds as well. Ask your local aqua-culture or fish pond supplier for information on Perox-Aid to learn more about this food fish production approved hydrogen peroxide based treatment.

89. Outdoor Fountains

Regular Strength to Extra Strength

For outdoor fountains not containing fish or plants, add one cup 3% hydrogen peroxide to the running fountain every few weeks to keep water clean and clear and algae and bacteria free.

If the fountain has a filtering pump, be sure to check the filter and clean it after adding the hydrogen peroxide to avoid clogging and pump failure.

For more serious cleaning jobs, you'll want to pick up some sodium percarbonate. To clean a fountain that has accumulated algae and organic matter start by draining the fountain. Then remove as much of the loose material as possible. Mix a paste of sodium percarbonate and hot water, being sure to wear gloves and generously coat the interior surfaces of the fountain. Let stand 20 minutes and then scrub with a stiff brush. Rinse thoroughly, and then refill with clean water.

Caution: sodium percarbonate dissolves in water to an approximate 27% hydrogen peroxide solution. This will burn your skin, and can kill fish and plants. If working in a garden area, be sure to continually run water over the area where the sodium percarbonate is draining to dilute it down to levels safe for the surrounding plants.

Auto

90. Window Cleaner

Regular Strength

As already noted in the indoor general cleaning section, hydrogen peroxide at 3% solution is an excellent glass cleaner. It is great for interior and exterior auto glass, leaving a clean fresh "no scent", removing musty odors, and quite literally streak free.

Automobile interiors can become quite grimy over time. Fingerprints, dirt, and the accumulations of dust, dander, pet and human skin oils can leave a filmy residue on interior glass surfaces. Hydrogen peroxide not only cuts through and cleans all that build-up; it also disinfects and freshens the interior of the automobile.

Generally we clean windows using spray bottle applications. In confined spaces, this can lead to spray splashing back towards you. Avoid getting peroxide in your eyes by either wearing safety glasses or reaching in with one arm to spray the glass while keeping your head outside the car.

91. Mirrors and Chrome

Regular Strength

If you think it's great on the interior windows, just wait till you see how well it works on your mirrors and chrome! Mirrors stay clear and chrome shines bright. That simple 3% solution spray bottle can handle it all.

92. Vinyl and Plastic Interior Surface Cleaner

Regular Strength

3% hydrogen peroxide spray is perfectly safe for use on vinyl, plastic and other man made interiors found in automobiles. It will deep clean, spot clean and deodorize the car all with one application. Simply spray down the interiors of car doors, moldings, dash units and glove boxes with the 3% solution and wipe clean with a sponge or clean damp cloth.

Use a dry rag to wipe down all the surfaces once they are clean to remove any excess moisture.

If you have fabric upholstery, test a small un-noticeable area first for color fastness, usually it's fine and there's no problem, but test to be sure. For deep-set stains, spray the hydrogen peroxide and let set.

You may wish to use a peroxide baking soda paste and let it completely dry, then vacuum up the powdery residue once it is fully dry. This will also do a terrific job of eliminating odors.

93. Carpet and Upholstery Spot Cleaning

Regular Strength to Double Strength

For carpeting stain removal in the car or truck, follow the same instructions we use for regular carpet and upholstery cleaning in the home. Saturate the area that is stained and let set for a few moments. Then blot. Repeat as needed.

Before doing stain removal of carpeting, it's a good idea to vacuum the carpeting and remove all loose dirt and debris. Then apply the hydrogen peroxide to the stain(s), let set for a moment, then blot and repeat as needed. Most of the time one application will do the trick. In fact, often you can watch the stains disappear right in front of your eyes as soon as the hydrogen peroxide is applied.

For tougher stains and/or serious odor and stain problems, mix a paste of 3% hydrogen peroxide and baking soda, lather on area generously, brushing into long nap to thoroughly saturate the base of the fibers, then let air-dry overnight. Remove with a vacuum with a brush end when completely dry. This will remove the odor as well as the stain and leave a fresh clean smell. Should you still have odor, the source of the odor is not the stained area you have just cleaned in this manner and you will have to figure out where the odor is coming from and then apply the peroxide and baking soda powder to the odor causing area.

Concrete Patios, Garage Floors and Driveways

94. Patio Cleaning

Regular Strength to Double Strength

Is your outdoor patio stained? Whether from mud, grease, or other contact with staining materials, concrete can become stained and unsightly over time. The peroxide based cleaners work best on organic stains: blood, grass, dirt, molds, mildews, fungi etc.

To remove stains from a concrete or brick patio, mix 2 ounces (~1/4 cup) sodium percarbonate to each gallon of hot (120 to 150 °F) water. Apply the liquid to the patio using a long handled scrub brush (non-metallic) stiff broom or mop. Let stand at least 15 to 30 minutes, and then rinse thoroughly with running water. In case of severe staining, scrub again before rinsing and repeat if needed.

You can leave a paste of sodium percarbonate and hot water in place on small spot stains for up to five hours, then scrub or remove and rinse. When using a paste, do not simply rinse off into grass or plant areas without first removing as much of the paste as possible unless at least five hours have passed to avoid burning of plants.

Sodium percarbonate is generally completely oxidized within approximately five hours of being mixed with hot water, and so the reactive processes have 'used up' the excess oxygen in the percarbonate, rendering it harmless to plants, pets and humans.

95. Garage Floors

Double Strength or
Extra Strength

Remove stains from concrete garage floors with the same formulation as for outdoor patios. Be sure to vacuum, sweep or otherwise remove all loose dirt, sawdust and other loose material before beginning. Use sawdust or clean sand to remove as much excess oil or gasoline from stains caused by these hydrocarbon compounds before attempting to remove the stains. This allows for proper disposal of the soiled sawdust or sand at the local hazardous materials recycling or drop off location in your local community and eliminates as much of these materials getting into the ground water around your home as possible.

Sprinkle the sodium percarbonate on the stained areas, brush vigorously with a non-metallic scrub brush dipped in hot water. Finally, rinse clean with flowing water.

(See cautions in Patio cleaning, above, for using concentrated sodium percarbonate.)

96. Driveways

Double Strength or
Extra Strength

Follow instructions for cleaning outdoor patios to clean your concrete driveway or apron. For larger stained areas, use a scrub brush on a pole. Be sure to run plenty of water when rinsing to avoid leaving residue of sodium percarbonate undiluted in nearby grass or garden areas.

Outdoor Decking

97. Wood Deck Stain Removal

Double Strength or
Extra Strength

Brighten wood and remove mildew, algae, nail head stains and other blemishes on outdoor wood decks. Mix 2 ounces (~1/4 cup) sodium percarbonate to each gallon of hot (120 to 150 °F) water. Use this solution in a large bucket with a mop, long handled scrub brush (non-metallic) or stiff broom. Apply the solution while scrubbing and let stand at least 30 minutes before rinsing thoroughly with running water.

For serious stains make a paste of the sodium percarbonate and apply with a stiff brush, let stand 30 minutes, and rinse with running water. Always use a 2nd application at the lower strength before moving to extra strength paste; and test for bleaching in an inconspicuous area first.

98. Wood Deck Cleaning

Regular Strength or
Double Strength

For general cleaning of wood decks a solution of sodium percarbonate at double strength (2 ounces per gallon of water) is sufficient to handle regular cleaning. It is easiest to apply using a bucket and long handled brush or stiff mop. Rinse and let air dry.

Regular cleaning will help prevent the build-up of algae and mildew, which can help prevent the onset of dry rot as well.

Vinyl Siding

99. Vinyl Siding General Cleaning

Regular Strength

Begin with a regular strength (1 ounce to one gallon water) solution of sodium percarbonate in warm to hot water. Remove all loose debris before washing using a spray hose or brush attachment.

Use a long handled brush, sponge or stiff mop to apply the sodium percarbonate from the highest point across and then work your way down the siding. Rinse with running water.

100. Vinyl Siding Stain Removal

Extra Strength

Tough mildew, mold or tree sap staining on vinyl siding can be removed using a small amount of sodium percarbonate paste applied directly to the stain and spread thinly over it. Let stand at least one hour, then scrub and rinse. Extremely tough stains may require two applications.

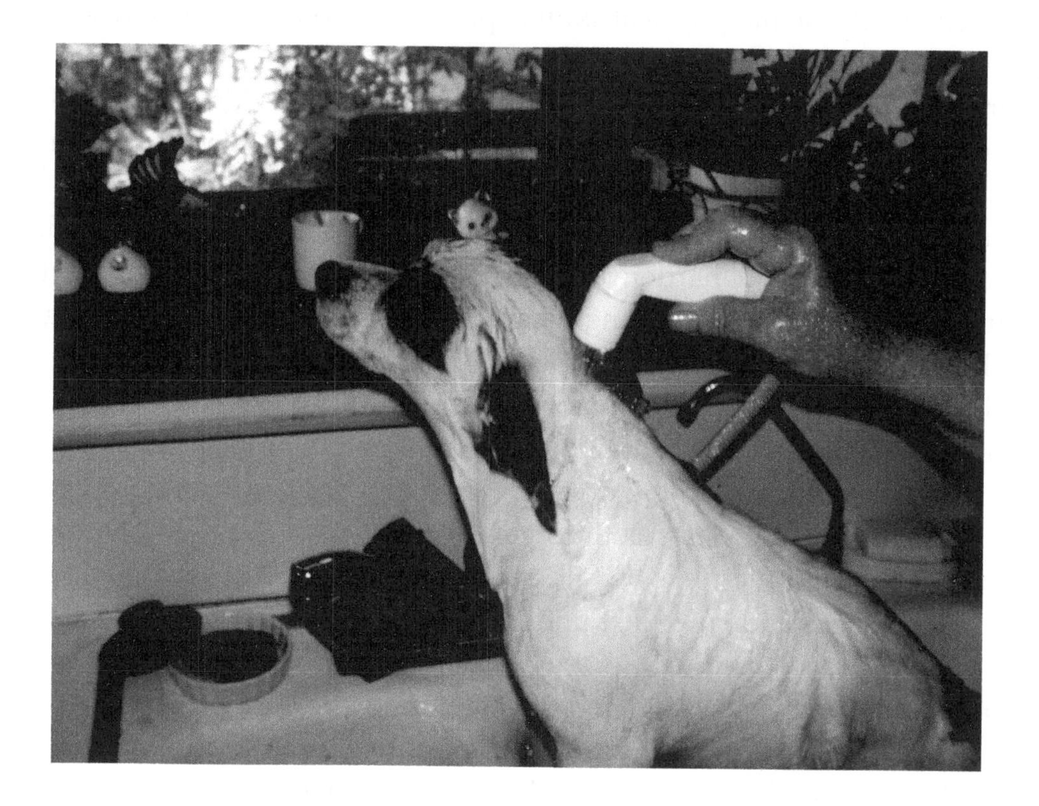

101. Skunk Odor Removal Recipe

Regular Strength

Perhaps the most amazing and potent use of hydrogen peroxide is as a skunk odor remover for dogs. While we have never actually used this remedy on a cat, (and have never had a cat sprayed by a skunk), chances are very good, if you could figure out how to get the cat to keep still for the treatment, that it would work equally well on cats as it does on dogs. Let's hope none of us ever has to find this out.

To treat a skunked dog, the first thing you'll need is some good latex or rubber gloves. You'll also want to put on your least favorite clothing. Some friends we know even wear a fish cutter's apron (one of those big yellow or white rubber coated things!) Depending on the size of the dog and the severity of the skunk encounter, you will have to make those decisions as you go.

The first thing to notice is how badly the dog has been sprayed. We have seen dogs who actually seem to love getting skunked, and in some cases the spray can be as heavy as to appear as a sticky clear coating on the dog's face, chest and head. Double yuck.

If this is the case, you will most likely need to apply the treatment twice: first to remove the actual physical skunk spray that is on the dog, and second to remove all residual odors.

For small dogs, mix 1 to 2 cups baking soda, a few drops of liquid dish detergent and 8 ounces of 3% hydrogen peroxide in a glass, plastic or ceramic container. A large bowl works best.

Continue to stir the mixture until it is a smooth consistent thick liquid. It will begin to settle and harden between stirrings, that's fine. Apply the mixture directly to the dog, starting where the spray is worst and working to the tail.

This can be tricky if the dog has been sprayed in the face. You do NOT want to get this mixture in or too close to the eyes. One way to solve this problem is to carefully wipe the dogs eye area first with a clean damp cloth and then use a separate clean cloth to gently cover their eyes so none of the solution can get into their eyes while you do the top of the head and areas around their cheeks and jaws (just under the ears and along the neck is often a very thickly sprayed area.)

Apply by stirring until liquid and then scooping one handful at a time and rubbing in to the dog's fur and skin. If the mixture gets too thick, simply add more hydrogen peroxide. If it gets too thin, add more baking soda.

Apply until you've coated the dog's fur from head to tail. This will take you 5 to 10 minutes. The longer the paste is on the dog, the better the results. However, it is a tough job to keep a dog coated in baking soda and hydrogen peroxide for any length of time! They want to shake (don't let them, it will spray little droplets of baking soda mix all over the place!) Most dogs don't like baths, so depending on your dog's temperament this job could be easier or more difficult to accomplish. Once the dog is thoroughly coated, begin to rinse off the paste starting at the tail and working your way back to the head. This lets the paste stay on the head and chest area longest where it is most needed.

When the dog is completely rinsed, check to see if you need to repeat the process. Unless the dog has sustained a direct hit to the face and front body area leaving visible spray on the fur, one application is usually sufficient.

➡ Note: Skunk spray is amazing stuff. It actually activates more strongly when in contact with moisture, so when you begin rinsing the dog you may suddenly smell the skunk odor again after the baking soda and hydrogen peroxide have already done most of their work. That's okay, just keep rinsing until you remove all the baking soda hydrogen peroxide mixture from the dogs fur.

From the moment you begin to apply the paste, the odor will become less unbearable, and by the time you are done with the "skunk odor bath" you will be greatly relieved.

The hardest thing about this treatment is that you actually have to bathe the dog… That is, you have to stand over this horrid smell up close and personal until the job is done. Don't torture yourself any more than necessary. Open nearby windows, do the bathing out of doors, or whatever it takes to relieve you of having to endure the up close experience of strong skunk odor.

This is absolutely the only method we have ever used in all our years of dog skunk encounters that actually works. In a funny confirmation of this fact, this actual skunk odor removal remedy was tested by the folks over at Myth Busters a couple of years back… and they too determined it to be the only genuine solution to skunk odor of the various methods they tried.

102. Emergency Remedy for Dog Poisoning

Regular Strength

Here is a potentially life-saving remedy for any dog that has eaten rat poison. First, it must be noted that if you only suspect rat poison ingestion and are not sure what the dog has eaten, it would be best to contact Pet Poison Control at (888) 426-4435 to describe symptoms and determine if this remedy is appropriate.

If you know your dog has eaten rat poison, immediately give the dog 2 teaspoons of 3% hydrogen peroxide orally. You may have to force feed and hold the mouth shut to force the dog to swallow. This will induce vomiting very quickly, getting the rat poison out of the dog.

We have friends who had to use this remedy when their dog ate rat poison at their next door neighbor's house, and he was fully recovered within 24 hours.

103. Litter Boxes

Regular Strength

Keep litter boxes fresh and clean with regular cleaning. A weekly spray down of the emptied litter box with a 3% hydrogen peroxide solution followed by a handful or so of baking soda, then wiped thoroughly with the mixture and discarded will keep the litter box smelling fresh and eliminate build-up of odors.

Every six months or so, give the litter box a complete disinfection and a fresh new start: completely empty the litter box, removing all solid material and fill with hot water mixed with 2 ounces (1/4 cup) of sodium percarbonate. Let stand at least 30 minutes, preferably up to 5 hours.

After soaking, scrub clean, drain and rinse and set to dry. The sodium percarbonate is great for septic tanks, plumbing pipes and toilet bowls, so flushing it down the drain is not a problem.

Between cleanings keep the litter box fresh by sprinkling a small amount of baking soda on the surface of the dry litter when changing the litter or adding fresh litter.

104. Ear Mite Prevention/Cure

Regular Strength

Does your dog have ear mites? Some dogs seem to be prone to them no matter how well cared for they are. If you notice your dog shaking his/her head often, it is wise to check for ear mites.

Use one to three cotton balls pinched together and gently swab out the inside of the ear. If it is covered with brown "crumbs" or comes up with a brown wet surface on the cotton ball chances are good your dog has ear mites.

To solve the problem, get some fresh clean cotton balls and saturate them with 3% hydrogen peroxide. Now gently swab the ear again, massaging the base of the ear while the cotton balls are in the ear. Remove, discard and repeat with clean dry cotton balls to remove the excess moisture. Regular applications of hydrogen peroxide to the ears will eliminate ear mites in dogs within weeks. Often one application is all it takes.

105. Wormer – Water Treatment

Regular Strength

Dogs can pick up worms just by walking around the neighborhood. They don't have shoes, and the most common worms can come in right through the pads of their feet.

The more dogs in the local area, the more likely yours are to pick up worms. Rather than spending lots of money on liquid wormers that the dogs simply refuse to take, throw up after you give them, or what have you, try adding a few drops of hydrogen peroxide to the water dish.

One teaspoon of 3% hydrogen peroxide per gallon of water every so often is a great defense against worms and infection for your dog. It is best to use this treatment sporadically, every few weeks or so, and not on an every day basis. Then add the hydrogen peroxide for several days running.

106. Aquariums

Regular Strength

Hydrogen Peroxide is used in large commercial aquariums around the world for algae control, oxygenating the water and general cleaning.

Ratios of hydrogen peroxide to aquarium water are extremely precise, and need to be tested carefully or the marine life in the aquariums could be threatened by improper applications.

For home aquarium use, it is best to remove any fish or living creatures before you first test using hydrogen peroxide in the aquarium. Amounts to be used are generally less than 1 ounce 3% solution per 100 gallons to start.

For home aquarium cleaning when the tank will be drained and refreshed, use sodium percarbonate in a soak to remove dead algae growth and stubborn stains. Use a regular strength solution, scrub as needed and rinse well.

Never add even a 3% hydrogen peroxide solution directly to the aquarium as it can harm the fish at such concentration. Instead, dilute at least 64 to one before adding small amounts (less than 8 ounces at a time) to the water.

Poultry & Livestock

107. Chickens

Regular Strength

The past decade has seen a rapid increase in the numbers of backyard chicken flocks in cities and towns across the United Kingdom and the U.S. Many cities have removed bans on backyard flocks. Although not so many metropolitan areas are ready to embrace roosters; hens, and particularly laying hens, are becoming a common urban phenomenon. Some even say the backyard chicken could be called the 'mascot' of the locally grown sustainable food movement.

Chickens are easy to raise and care for, and fresh eggs provide incentive enough for many backyard gardeners to bring chickens into the mix.

At the same time, in the more commercially focused organic farming and sustainably focused farms, free range and pasture raised chickens are also on the rise. And along with the rise in sustainably raised animals for meat, and the demand for grass fed meats and poultry, is a new rising trend in agriculture: the use of hydrogen peroxide based solutions for water purification and in the drinking water systems of livestock and poultry.

It is worthy of note that one of the most famous accounts of healing birds, "The Bird Man of Alcatraz" is a true account of using hydrogen peroxide to cure bird (or 'avian') flu. So perhaps it is not surprising that those seeking healthy, safe and effective and environmentally sustainable solutions for bird and animal care in farming are turning to hydrogen peroxide.

Less well known but no less significant, is the evidence from small numbers of farmers in Pennsylvania who successfully protected their flocks of poultry from an H5N1 bird flu outbreak which began in 1983. Millions of birds were exterminated during the course of the government program to contain the outbreak. These farmers, supplementing their poultry watering systems with hydrogen peroxide were able to effectively avoid contamination of their flocks by the disease and therefore were able to avoid the loss of their flocks to mandatory extermination due to infection.

In our own backyard flock, we simply add about 20 drops of 35% food grade hydrogen peroxide to each one half gallon fresh watering dish as we put out clean water every two or three days.

The rough calculation we shoot for is approximately a .075% solution of hydrogen peroxide. It takes very little peroxide to do the job of keeping birds healthy. Many farmers report that birds are less aggressive and easier going when peroxide is added to their drinking water systems.

108. Livestock

Regular Strength

Starting nearly 20 years ago, the use of hydrogen peroxide based solutions in agriculture has expanded significantly. In fact, a 2007 FDA approval of Perox-Aid is the first approval of any remedy for aqua-culture fish and is a food grade hydrogen peroxide based product. At the same time, the use of hydrogen peroxide in livestock farm operations is also on the rise.

One company, Essential Water Solutions, Inc. has made great strides in the areas of water purification with hydrogen peroxide for agriculture, livestock and rural water treatment. You can learn more about their products and applications at http://foodgradeh2o2.com/essential-water-solutions. You can find direct testimonials of pig farmers, cattle ranchers and many others who have been successfully using their specialized peroxide water delivery systems for livestock over the past 15 years.

Distribution of their flag ship product, Oxy-blast has proven remarkably successful with poultry, pigs, cows and other livestock. This hydrogen peroxide metered dosing systems in watering systems for animals, has demonstrated such evidence as marked reductions in stillbirths, increased health and vigor in new litters, generally improved health as well as either a drastic reduction in, or a complete elimination of the need for antibiotics and other pharmaceutical medication of the animals.

109. Wood Refinishing

Extra Strength

Want to strip that old dresser or bookcase and refinish it? The next time you need to strip some furniture, try using 35% hydrogen peroxide instead of toxic furniture finish strippers.

You will still need your rubber gloves because 35% hydrogen peroxide will burn your skin on direct contact. Although it will do no permanent damage, it is quite uncomfortable. Should you accidentally spill some 35% hydrogen peroxide on your skin just flush with fast running water for at least 3 minutes. It may still sting a bit, and the skin may turn white temporarily.

You can apply the 35% hydrogen peroxide with a rag or by pouring it directly on the furniture surface and letting it sit as it begins to oxidize and loosen the old finish.

Use your scraper blade to remove the loosened finish. Repeat the process until you have removed most of the finish, and then wipe the surfaces dry. Once the wood surface is completely dry again, you can proceed to the sander to take off the last rough bits and get your furniture ready for its new finish.

If you don't have any 35% hydrogen peroxide, you can also use sodium percarbonate for this job. Simply add hot water to create a full strength paste of sodium percarbonate, and smear it over the furniture surface. Wait an hour, testing occasionally to see if the finish is ready to be scraped off, then scrape off the old finish, rinse off any residual sodium percarbonate with a damp cloth or sponge, and wipe dry. Let the piece stand until it is completely dry before proceeding to sanding.

110. Carbon and Grease Buildup Stripping

Double Strength

To remove build-up of carbon from barbeque grilling racks, soot stains on fireplace bricks or any place where cooking with fire or heating with fire has left soot and staining, make a paste of sodium percarbonate and hot water.

Apply the paste to the affected areas and let stand at least one hour. Scrub and rinse using hot water. For very stubborn carbon deposits you may need to soak the item overnight in a strong cleaning solution (3 to 4 ounces sodium percarbonate per gallon of hot water) and then scrub and rinse.

111. Oil Rags, Work Clothes Presoak

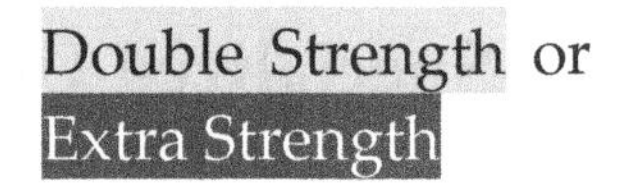

See Laundry Heavily Soiled Items.

112. Surfers Sinus

This newest inclusion comes directly from our family experience. With more than a few surfers in the family, it was inevitable someone would develop sinus pain after a serious wipeout. Whether it's serious pain or simply swelling and congestion, hydrogen peroxide is a particularly excellent cure. When it comes to sinus pain; there is nothing that makes you want to crawl right out of your skin like an ongoing sinus headache that will not quit.

The recipe is quite distinct and should be followed carefully and explicitly. No short cuts. No tap water. No fudging! For complete instructions read http://foodgradeh2o2.com/sinus-flush.

Appendices

Appendix A: At A Glance: Basic Cleaning Product Replacement Guide

Ammonia General Household Cleaners, Spray Cleaners, Floor Cleaners	3% regular strength hydrogen peroxide spray OR general cleaning solution of 1 oz. (~1/8 cup) sodium percarbonate to one gallon warm/hot water.
Oven Cleaner	Paste of sodium percarbonate and hot water applied for 10 to 30 minutes, then wipe and rinse clean.
Bathroom Tile Grout Cleaner	Sodium percarbonate paste mix, apply and let stand for up to 1 hour, scrub if necessary, rinse clean.
Carpet Spot Stain Remover	3% hydrogen peroxide spray solution, saturate, blot and repeat as needed.
Carpet Deep Stain and Odor Removers	Paste of 3% hydrogen peroxide and baking soda, apply, brush in, and allow to stand until dry. Then vacuum up powder.
Carpet Cleaning – General	Add 1 ounce sodium percarbonate or 4 to 8 ounces of 3% hydrogen peroxide to 1 gallon of shampooing tank solution.
Kitchen Disinfection/Sterilization Cutting Boards, Counter Tops and Raw Food Handling Surfaces and Utensils	Spray with 3% hydrogen peroxide followed by a spray of vinegar. Wipe or rinse clean.
Window Cleaner	3% hydrogen peroxide solution in a spray bottle.
Laundry Pre-Soak	1 to 3 ounces sodium percarbonate to 1 gallon of hot water depending how soiled items to be pre-soaked are.
Fine Fabric Spot and Stain Removal	Fine fabrics: 3% hydrogen peroxide solution spray (you can also use a Q-Tip to apply to small spot stains.)

Cement Cleaners	General Cleaning – 2 ounces sodium percarbonate 1 gallon hot water, apply, scrub and let stand 15 to 30 minutes, rinse thoroughly.
Pre Wash for Indoor Painting	3% hydrogen peroxide solution in spray bottle, spray and wipe clean to remove grease, dirt, and debris before painting.

Appendix B: Cleaning Formulations and Home Cleaning Recipes

- **General Cleaning Solution:** 1 ounce sodium percarbonate (1/8 cup) to 1 gallon hot water OR 3 percent hydrogen peroxide in spray bottle
- **Odor Remover for pets, carpets, upholstery:** 1 to 2 cups baking soda, 6 to 8 ounces 3% hydrogen peroxide, 3 to 6 drops liquid dish detergent.
- **Deep cleaning stain removal and scouring paste:** mix sodium percarbonate with only enough water to make a paste. (Wear gloves and spot test for bleaching). OR Use 3% hydrogen peroxide spray directly on stains and spots. 3% liquid can be used to saturate deep stains without worry of bleaching. Mix with baking soda to make scouring paste.
- **Kitchen Disinfection/Sterilization:** Spray bottle of 3% hydrogen peroxide solution and spray bottle of acetic acid (vinegar) use consecutively, the order is not important. Keep the two liquids in separate spray bottles to retain shelf life and effectiveness. This same sterilization duo works great in the bathroom as well.

Appendix C: Hydrogen Peroxide Dilution Tables

8% H_2O_2 Dilution Table

Percentage of Desired Solution	Amount of 8% H_2O_2 [units]	Water to Add [units]
1.00%	1	7.00
1.50%	1	4.33
2.00%	1	3.00
3.00%	1	1.67
4.00%	1	1.00
5.00%	1	0.60
6.00%	1	0.33
7.00%	1	0.14
8.00%	1	0.00

Example: To make a 3% H_2O_2 solution out of 5 oz of 8% H_2O_2 add 5*1.67=8.35 oz of distilled water.

12% H_2O_2 Dilution Table

Percentage of Desired Solution	Amount of 12% H_2O_2 [units]	Water to Add [units]
1.00%	1	11.00
1.50%	1	7.00
2.00%	1	5.00
3.00%	1	3.00
4.00%	1	2.00
5.00%	1	1.40
6.00%	1	1.00
7.00%	1	0.71
8.00%	1	0.50
9.00%	1	0.33
10.00%	1	0.20
11.00%	1	0.09
12.00%	1	0.00

Example: To make a 3% H_2O_2 solution out of 5 oz of 12% H_2O_2 add 5*3=15 oz of distilled water.

35% H_2O_2 Dilution Table

Percentage of Desired Solution	Amount of 35% H_2O_2 [units]	Water to Add [units]
1.00%	1	34.00
1.50%	1	22.33
2.00%	1	16.50
3.00%	1	10.67
4.00%	1	7.75
5.00%	1	6.00
6.00%	1	4.83
7.00%	1	4.00
8.00%	1	3.38
9.00%	1	2.89
10.00%	1	2.50
11.00%	1	2.18
12.00%	1	1.92
13.00%	1	1.69
14.00%	1	1.50
15.00%	1	1.33
16.00%	1	1.19
17.00%	1	1.06
18.00%	1	0.94
19.00%	1	0.84
20.00%	1	0.75
21.00%	1	0.67
22.00%	1	0.59
23.00%	1	0.52
24.00%	1	0.46
25.00%	1	0.40
26.00%	1	0.35
27.00%	1	0.30
28.00%	1	0.25
29.00%	1	0.21
30.00%	1	0.17
31.00%	1	0.13
32.00%	1	0.09
33.00%	1	0.06
34.00%	1	0.03
35.00%	1	0.00

Example: To make a 3% H_2O_2 solution out of 5 oz of 35% H_2O_2 add 5*10.67=53.35 oz of distilled water.

Mixing Ratios for Sodium Percarbonate

- **General Cleaning:**
  Mix 1 dry ounce by weight of percarbonate in a gallon of warm or hot water.
  1 ounce dry weight is ~ 1/8 cup dry measurement; 2 ounces is ~ 1/4 cup dry measure.
- **Heavy Cleaning/Stain Removal:**
  Start with general cleaning solution first and if more strength is needed, increase dry percarbonate to 2 ounces (or 1/4 cup dry measure).
- **Soaks:**
  Pre-soak with 1 ounce (1/8 cup dry measure) per gallon of warm to hot water for one hour before washing.
- **Paste:**
  Mix 2 dry ounces (1/4 cup dry measure) of sodium percarbonate with 2 tablespoons of hot water.

> ➠ NOTE: Always wear gloves and eye protection when working with the paste strength of sodium percarbonate as it is roughly equivalent to a 27% peroxide solution in oxidizing power!

Appendix D: Hydrogen Peroxide Stabilizers in the Marketplace

Hydrogen peroxide is fairly stable in concentrations of up to 40 percent. It does, however, break down with exposure to water, sunlight, and over time, into water and oxygen. For this reason it is often used in high level aquatic applications to introduce additional oxygen to the water in large aquariums and closed system water environments.

However, hydrogen peroxide that is produced for the technical industry, as well as over the counter drug store hydrogen peroxide, is all treated with a number of possible stabilizing agents to retard dissociation of the hydrogen peroxide into its water and oxygen components. The most common stabilizing agents include "Acetamilide", "phenol", "tin", "Colloidal stannate", "sodium pyrophosphate" (present at 25 - 250 mg/L) and "organophosphonates", "nitrate".

Most of these stabilizing agents are toxic to humans. They are used in hydrogen peroxide stabilizing for specific applications, none of which include regular household cleaning, bathing, food preparation etc. For this reason it simply makes sense to avoid use of these grades of hydrogen peroxide around the home.

Let's have a look at an excerpt from Wikipedia regarding the use and applications of this "stabilizing agent" and see if you can figure out why any personal care, food related or in home cleaning application of hydrogen peroxide is better off without it!

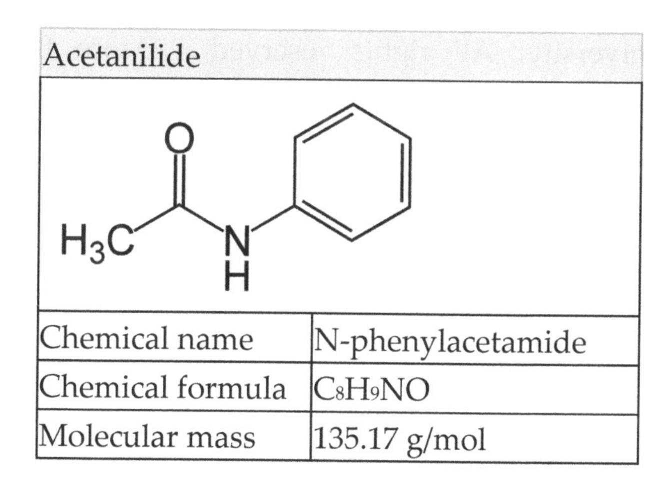

Acetanilide	
Chemical name	N-phenylacetamide
Chemical formula	C_8H_9NO
Molecular mass	135.17 g/mol

Applications

Acetanilide is used as an inhibitor in hydrogen peroxide and is used to stabilize cellulose ester varnishes. It has also found uses in the intermediation in rubber accelerator synthesis, dyes and dye intermediate synthesis, and camphor synthesis. Acetanilide was used as a precursor in penicillin synthesis and other pharmaceuticals and its intermediates.

Acetanilide has analgesic and fever-reducing properties; it is in the same class of drugs as acetaminophen or paracetamol. Under the name acetanilid it formerly figured in the formula of a number of patent medicines and over the counter drugs. In 1948, Julius Axelrod and Bernard Brodie discovered that acetanilide is much more toxic in these applications than other drugs, causing methemoglobinemia and ultimately doing damage to the liver and kidneys. As such, acetanilide has largely been replaced by less toxic drugs.

In the 19th century it was one of a large number of compounds used as experimental photographic developers.

Appendix E: Resources

Visit the Resources Page at FoodGradeH2O2.com:

http://foodgradeh2o2.com/resources/

Buying Food Grade H_2O_2 at 35%, 12%, 8%, and 3% Solutions. Here are the best prices available in the market according to my research as of 2013:

4 Quarts certified 35% Food Grade Hydrogen Peroxide:
http://foodgradeh2o2.com/4-quarts-35pct-fgh2o2

Food Grade H_2O_2 12% 16oz.:
http://foodgradeh2o2.com/16oz-12pct-fgh2o2

Food Grade H_2O_2 8% 1 Pint:
http://foodgradeh2o2.com/1-pint-8pct-fgh2o2

Case Food Grade H_2O_2 8% 9 Pints:
http://foodgradeh2o2.com/9-pints-8pct-fgh2o2

Food Grade H2O2 3% 16oz:
http://foodgradeh2o2.com/16oz-3pct-fgh2o2

Interestingly, the least expensive Sodium percarbonate sold online right now is Oxiclean! Find it here: Sodium Percarbonate - Oxiclean 5lb bucket:
http://foodgradeh2o2.com/5lbs-oxiclean

Sustainable green products education site debunking false claims of green products – The Sins of Green Washing:
http://sinsofgreenwashing.org/

Hydrogen Peroxide Science Related Resources:

- NSF (National Science Foundation)
  http://foodgradeh2o2.com/nsf

- Pure Energy Systems Wiki – Community-built energy information site
  http://foodgradeh2o2.com/peswiki

- Dr. Majid Ali's Aging Healthfully Virtual Library - Dr. Majid Ali-Integrative Medicine Editor of Integrative Medicine journal
  http://foodgradeh2o2.com/majidali

- The Many Benefits Of Hydrogen Peroxide - Articles
  http://foodgradeh2o2.com/majidali-h2o2-articles

About the Author:

An avid enthusiast of all things in nature since childhood, Ms. Mundt has spent a life time appreciating the wonders of the natural world. Her love of nature led her to seek solutions for housekeeping and homemaking which were "earth friendly" and "pets and humans friendly" as well.

Over the course of raising four boys, she has experimented with hundreds of different cleaning, laundering, bleaching, stain and odor removal and other household products and remedies. Not satisfied to "go green" without also "staying clean" she has been a tough customer for many so-called "green" products, demanding they do the job as well or better than their not so healthy or safe chemical competitors.

In recent years Ms. Mundt has turned her attention to research and writing in the areas of natural living, sustainable practices and inner exploration. She currently makes her home in Oregon, with her husband, 2 dogs, 6 chickens and 3 ducks.

Lightning Source UK Ltd.
Milton Keynes UK
UKOW06f1000161114

241676UK00010B/168/P

9 781630 221980